AUTOMATIC
FOR THE CITY

DESIGNING FOR PEOPLE IN
THE AGE OF THE DRIVERLESS CAR

Riccardo Bobisse and Andrea Pavia

RIBA ⢨⢨⢨ **Publishing**

© RIBA Publishing, 2019

Published by RIBA Publishing, 66 Portland Place, London, W1B 1NT

ISBN 978-1-85946-861-6

British Library Cataloguing-in-Publication Data
A catalogue record for this book is available from the British Library.

Commissioning Editor: Alexander White
Production: Richard Blackburn
Designed and Typeset by Studio Kalinka
Printed and bound by Westdale Press, Cardiff
Cover illustration: Giordano Poloni / agencyrush.com

While every effort has been made to check the accuracy and quality of the information given in this publication, neither the Author nor the Publisher accept any responsibility for the subsequent use of this information, for any errors or omissions that it may contain, or for any misunderstandings arising from it.

www.ribapublishing.com

ACKNOWLEDGEMENTS

To all our families, the ones that nurtured us yesterday and the ones we are tendering today, and to our patient wives Teodora and Holly, who supported us while developing this project.

We would also like to thank our publisher, the RIBA and Alex White in particular, as well as Mike Davies and Andres Sevtsuk for their thoughtful contribution.

A special thanks to Steer, where we initiated our research into the topic of CAVs and cities and which constitutes the genesis of this book, and to all our colleagues and ex-colleagues around the world.

CONTENTS

PART 1
CARS, PEOPLE AND URBAN DESIGN

PREFACE   vi

ABOUT THE AUTHOURS   ix

1

CARS AS MAKERS AND DESTROYERS OF SPACE   3

ENTER THE CAR   4
PEOPLE AND CARS: A LOVE AFFAIR   10
CAR RESISTANCE   21
THE RISE OF AUTOMATION   25

2

SETTING THE SCENE   27

THE AUTONOMOUS VEHICLE   28
A PLAUSIBLE SCENARIO   38

3

CAVS PRINCIPLES FOR PEOPLE-CENTRIC URBAN DESIGN   43

DESIGN STREETS NOT ROADS   44
KEEP IT LEGIBLE   46
SHARE CAVS, SHARE STREETS   47
REALLOCATE SPACE   48
PHASE OUT CARS   49
ENABLE NEW ARCHITECTURE   50
MAKE IT RESILIENT   52
WIN FAST (WHERE THE OPPORTUNITY ALLOWS)   55

PART 2
VISIONS FOR THE FUTURE

4
LONDON VS LOS ANGELES: 59
TWO CASE STUDIES FOR
TWO VISIONS
LONDON IN CONTEXT   61
LOS ANGELES IN CONTEXT   63
SITES SELECTION CRITERIA   65

5
LONDON   67
PRACTICAL STUDY AREA: 68
SHOREDITCH
SPACE BATTLES   71
CELLS AND CORRIDORS   74
THE FUTURE STREETS   77
OF SHOREDITCH
MAIN ROADS   77
C-CORRIDORS   79
LOCAL STREETS   82
STREET TYPE RELATIONSHIPS   85
NEW TECHNOLOGY,   85
NEW BUILDINGS
FURTHER IMPLICATIONS FOR   86
THE BUILT FORM
WHAT IF CARS WERE BANNED   88
FROM LONDON'S STREETS?

6
LOS ANGELES   91
PRACTICAL STUDY AREA: 92
DOWNTOWN LOS ANGELES
CAVS SUPERGRID   98
A RESILIENT URBAN FORM   102
MOVEMENT AND PLACE   105
PAVEMENT KERB   111
CAVS MOBILITY HUBS (NEMOHs)   112
PARKING AND OTHER   113
IMPLICATIONS FOR THE
BUILT FORM
INTERIM PHASE   114

7
A WAY FORWARD   117
CAVS AND THE URBAN   119
ENVIRONMENT
IMPLEMENTATION AND   121
RECOMMENDATIONS
FOR CITY MAKERS

APPENDICES
I CAVS AND URBAN DESIGN   125
BY MIKE DAVIES
II AN INTERVIEW WITH ANDRES   129
SEVTSUK

CAVS GLOSSARY FOR DESIGNERS   137
BIBLIOGRAPHY AND   142
FURTHER READING
REFERENCES   145
INDEX   149
IMAGE CREDITS   150

PREFACE

The analogy of the city as human body has endured since the Renaissance, sustained by designers and theorists like Leon Battista Alberti or Francesco di Giorgio Martini to give sense and structure to the city's different parts, functions and interrelations.

Using this analogy we imagine the city's mobility system as a human skeleton, providing support, movement, and regulation to the other parts of the body such as muscles and organs. As technologies for urban mobility evolve, so too does the body. With the revolution of the private automobile after World War I, for example, the world witnessed a rapid and unprecedented transformation of cities – the body mutated beyond recognition. Today, we are on the verge of a similar revolution. According to the latest industry predictions, connected autonomous vehicles (CAVs) will become a common feature in the wealthiest urban environments within the next 10 years.[1] How will this shift change the form of our cities? And how will this evolution unfold? Is the body of the city going to mutate dramatically once again and, if so, what will this look like? Or is the analogy shifting altogether from the human body to the complexity of the human brain?

This book speculates about the ways in which mobility and technological changes may affect the urban form of cities in the next 20 years. New technologies have the potential to radically change cities' transport systems and improve their efficiency, cost, and inclusiveness.

We deliberately avoid developing forecasts, instead emphasising possible scenarios based on the information available today. CAVs and cars will initially coexist and both will require space. Cities must ensure a smooth transition to the new technology, and will have to do so without compromising the potential benefits of CAVs through the ever-pressing necessities of the current car-centric system.

This book focuses on two areas in the central urban cores of Los Angeles and London in case studies that provide sounding boards for proposed approaches by imagining how they would function with CAVs. In fact, the two cities are emblematic within Western societies and display a rich array of the issues faced by the metropolis worldwide. The car paradigm was applied to both, but while relatively young Los Angeles proved freer in adapting to the needs of the car, old London's physical constraints led to a different outcome. In both cases, the effect of cars on urban form was radical. Once again, urban environments in Los Angeles and London are at the forefront of a mobility revolution that will eventually unfold globally.

As urbanists, we appreciate the benefits that this technological revolution may bring in terms of sustainability, liveability, and accessibility, and welcome this leap forward. However, we also value the lessons learnt from the mistakes of the last century and recognise that technology is a tool, not a means. As designers, we want to contribute to the debate and help urban dwellers to envision and capitalise on the transformations that this revolution will bring. The central idea presented in this book is that cities cannot afford to be shaped once again by a new technology that will create a new set of infrastructural rigidities. Urban populations will continue to increase exponentially, and CAV technology will significantly disrupt the business-as-usual model. Mobility, congestion, pollution and sprawl will be among the major challenges to cities, and tackling these issues will be (and already is) a priority for regional and local policy makers. On this basis, the book identifies a series of design ideas and recommendations for stakeholders involved in city-making, and discusses how best to address, prepare for and exploit these imminent changes while acknowledging that human scale and well-being will have to remain the fundamental driving principles of urban design.

This book is not intended as an exercise in futurology but as a platform for discussion. So far, the topic of driverless cars has seen little attention from urban designers, architects, planners, politicians and citizens and has mainly focused on the end state, when CAVs will be the only driving technology. Urban designers cannot afford to delay engaging in the debate if they want to keep improving cities and avoid falling into the trap of 'simplification' for the sake of accelerating adoption. Making the urban environment predictable to enable a new technology would, in fact, end up denying the city's essence and eroding its quality as a place of exchange, unpredictability and surprise.

ABOUT THE AUTHORS

RICCARDO BOBISSE

Riccardo Bobisse is a practicing urbanist based in London with 15 years of professional experience in Europe and overseas. He specialises in masterplanning with a specific focus on mixed-use urban schemes and town centre revitalisations strategies. Riccardo holds multiple academic degrees in urban design, planning and regeneration from Westminster University, University College London's Bartlett, and Venice University (IUAV) and has built his professional experience in both the public and private sector. He is also a design review panelist for Design Council/CABE.

https://uk.linkedin.com/in/riccardo-bobisse

ANDREA PAVIA

Andrea Pavia, AICP is an urban designer with almost twenty years of professional experience in award-winning projects in the USA, Europe, China and the Middle East. Andrea's research focuses on the issues of ecology, energy and mobility as drivers in the design of urban frameworks and at the centrality of public spaces in city making. His projects and articles have been featured in Urban Land, American Planning Association - Los Angeles and in PianoProgettoCitta' among others. His essay on "Walking as a luxury activity" has been recently published by Routledge. Andrea graduated from the Harvard GSD and he has been adjunct instructor in urban design at the Boston Architectural College.

https://www.linkedin.com/in/andreapavia

3.30   3.30   3.30   3.60

5m

PART 1

||||||||||||||||||||||||||||||||||||||||||||||||||||||||||||||||||||||||||

DESIGN, PEOPLE
AND AUTOMOBILES

It is a powerful symbol, showing that a citizen on
a thirty-dollar bicycle is as important as one in
a thirty thousand dollar car. A protected bicycle
lane along every street is not a cute architectural
fixture, but a basic democratic right, unless one
believes that only those with access to a car have
a right to safe mobility.

Enrique Peñalosa, 'Politics, Power, Cities,' in The Endless City.
Phaidon Press, 2008

3.30

27.5m

Chapter 1

CARS AS MAKERS
AND DESTROYERS OF SPACE

The cities will be part of the country; I shall live 30 miles from my office in one direction, under a pine tree; my secretary will live 30 miles away from it too, in the other direction, under another pine tree. We shall both have our own car. We shall use up tires, wear out road surfaces and gears, consume oil and gasoline. All of which will necessitate a great deal of work… enough for all.

Le Corbusier, *The Radiant City*, 1933

Transport innovations have always structured the making of cities but never to the extent brought about first by trains and later by cars. This is particularly evident in analysing the evolution of Western cities from approximately the mid–19th century to today.

ENTER THE CAR

Science fiction author Arthur C. Clarke observed that the world has moved on wheels for 6,000 years in an unbroken sequence from the ox cart to the Rolls-Royce and Mercedes-Benz.[1] Yet the impact that a disruptive transportation technology like the automobile has brought to urbanisation and lifestyles is a very recent one. Prior to the appearance of the automobile, horses, walking and streetcars were the major modes of travel within cities, and like a city's skeleton provided support to the other parts of the city-body.

Walking cities

An interpretation of how cities evolved from walking cities to automobile cities from a transport point of view, and of how an elaborate and powerful 'car culture' developed around greater dependence on automobiles, has been provided by Preston, Schiller and Kenworthy.[2] In their work they identify three types of city: walking cities, transit cities and automobile cities. Cities were mostly dependent upon walking for their circulation needs for most of their existence. In Europe, the walking city was dominant until 1850. Other slow transport modes like horse-drawn or waterways travel were partially available to the masses but in order for the city to remain accessible all destinations had to be within about half an hour's walk, travelling at about 5 km/h. This meant that urban environments remained for the most part small and dense compact nuclei with a concentrated shape, and with highly mixed land uses.

Transit cities

From the 19th century, railways introduced radical changes to land uses, employment patterns, social interactions, infrastructure and the distribution of goods. At the end of the 17th century most Londoners travelled to work and to shops and markets on foot. By the beginning of the 20th century, the expansion of the metropolis meant that thousands commuted daily from the suburbs by omnibus, tram, railway and even steamboat.[3] Transit cities started developing in the new industrial world around 1850 with the advent of new transport technologies like the steam train and the electric tram. These modes facilitated faster travel, with an average jump from 5 km/h to 15 km/h, and allowed for the expansion and realisation of larger cities within fixed development corridors, with smaller satellite communities linked through transit routes to a central nuclei. Most new developments still had to be within walking or cycling distance from public transport to provide access to the larger pool of jobs and services in the city.

FIGURE 1.1
LONDON GROWTH RINGS. THE CIRCLES ARE AN APPROXIMATION OF THE BUILT-UP AREA. AFTER WWII THE CAPITAL CONTINUED TO EXPAND SPATIALLY WHILE THE OVERALL POPULATION REMAINED STATIC, RESULTING IN LOWER DENSITIES IN FACT, LONDON HAS NEVER EXCEEDED ITS PEAK POPULATION OF 1939[4]

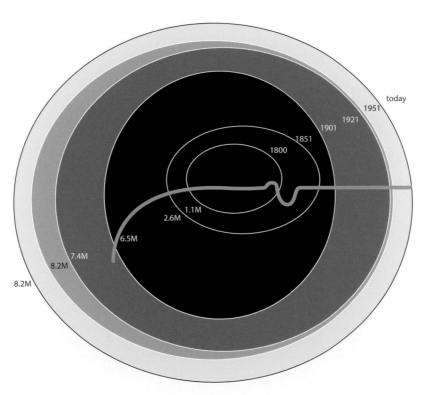

Therefore, these cities still presented a high density of mixed-uses and a well-defined edge to urban development. In a world touched by the industrial revolution, particularly in the West, this was according to Preston, Schiller and Kenworthy the dominant type of urban development pattern from approximately 1850 to 1940.

Automobile cities

The first private car appeared on the streets of London in the 1890s, while the first gasoline-powered vehicle appeared on the streets of Los Angeles in 1897. Worldwide, the arrival of the car facilitated the uninhibited outward expansion of cities. The automobile's speed allowed the city to get much bigger and densities of development dropped dramatically. Thanks to allowing easier access to remote places, the automobile quickly became the most popular mode of transport in developed countries. Passenger cars emerged as the primary means of family transportation, with an estimated 0.75 billion in operation worldwide today. In most North American cities today, the automobile has a typical modal share of all daily trips in the range of 80–95 per cent[5], with a quarter of all automobiles on the planet being in the United States.[6]

The automobile revolution began around 1890 when car production started rising and investors in both Europe and North America began to back various vehicle production technologies. In the early 20th century, once the initial unreliability issues were solved, gasoline-powered cars entered mass production and emerged as the dominant technology. In 1913 the Ford Model T, created by the Ford Motor Company five years earlier,

became the first automobile to be mass produced on a moving assembly line. Henry Ford produced eight models of car before the Model T with which his name became synonymous. By 1927 Ford had produced over 15 million Model T automobiles. In 1904 1,600 motor vehicles cruised the streets of Los Angeles and by 1915 Los Angeles County counted 55,217 motor vehicles, leading the world in per capita ownership of automobiles as it continues to do today.[7] By the mid-1920s the American internal combustion engine automobile had won the revolution that Ford had begun, and the manufacture and sale of automobiles had become an important component in the American economy.

Ultimately, the automobile made regular medium-distance travel more convenient and affordable, especially in areas without railways. Cars did not require rest, they were faster than horse-drawn conveyances, and soon had a lower total cost of ownership. The new product for mass mobility was successful for a number of reasons: it provided a streamlined journey from beginning to end; a safe environment, protection against the weather and unfamiliar faces and people; and 'a shield against the insecurity of the outside world'.[8] Since the beginning of the 20th century, Modernist movements had embraced and promoted the car, from the Italian Futurists' celebration of speed to Le Corbusier's 'road monopolised by cars'.

It is ultimately with mass production that the automobile's impact on existing and planned city forms became significant. Beginning in the 1940s, most urban environments in the United States – thanks to pro-car policies – lost their streetcars, cable cars, and other forms of light rail to diesel-burning motor coaches and buses. The growing importance of the car contributed decisively to changes in land-use patterns, employment distribution, shopping habits, social interactions and manufacturing priorities, affecting the way that cities were planned. The construction of highways and the Federal-Aid Highway Act of 1956, paired with urban renewal policies, further accelerated this process to the point of urban flight. In the decades following World War II, the automobile united with the single-family dwelling to create suburbs. Community standards of the past, driven by scarcity and the need to share public resources, gave way to new credos of self exploration. As the US economy of the 1950s and 60s boomed, car sales grew steadily from 6 million units per year to 10 million. Married women also entered into the economy and two-car households became commonplace.

In addition to public money for roadway construction, car use was also encouraged in many places through new zoning laws that required any new business to construct a certain amount of parking based on the size and type of its facility. The effect was to create many free parking spaces, and to set business spaces further back from the road, leading to more open settlements that made a carless lifestyle increasingly unattractive, or even impossible. In addition many new shopping centres and suburbs did not build pavements, making pedestrian access dangerous. This had the effect of encouraging people to drive, even for short, walkable trips, thus increasing and solidifying American auto-dependency. As a result of this change, employment opportunities for people who were not wealthy enough to own a car or who could not drive due to age or physical disabilities became severely limited. The loss of pedestrian-scale villages also disconnected communities.

The use of cars for transportation started creating barriers by reducing or eliminating open space dedicated to walking and cycling. Initially minor nuances, incrementally over time these barriers became the dominant structure of entire urban landscapes, physically blocking any pedestrian or cycling movement, segregating communities and becoming a major threat to the safety of children and the elderly. Car infrastructure became a major land use allocation, leaving less land available for other purposes. It was the rise of the Automobile City, of cities that facilitate and encourage the movement of people, via private transportation, through 'physical planning' such as built environment innovations (street networks, parking spaces, automobile/pedestrian interface technologies and low-density urbanised areas containing detached dwellings with driveways or garages) and 'soft programming' such as social policy surrounding city street usage (traffic safety/automobile campaigns, automobile laws and the social reconstruction of streets as reserved public spaces for the automobile).[9]

In many countries, such as the United States, the infrastructure that made car use possible - highways, roads and parking lots, was funded by government and supported through zoning and construction requirements.[10] Zoning laws in many areas required that large free parking lots accompany any new buildings, while municipal parking lots were often free or did not charge a market rate. Hence, the cost of driving a car was subsidised, supported by business and government, which covered the cost of roads and parking. This was in addition to other externalities car users didn't have to pay for such as accidents or pollution.[11]

FIGURE 1.2
SUBURBAN DEVELOPMENT PATTERNS IN
AUSTIN, TEXAS

PEOPLE AND CARS:
A LOVE AFFAIR

The post-war experience of the automobile is perhaps the purest synthesis of national status and identity with a particular consumer product, a love affair largely unsullied by the harsh realities of environmental protest, high petrol prices and choked highways. Then, as now, automobile ownership was aspirational.

Bell, J, Carchitecture, 2001

Whether you are watching TV, a 5-second YouTube advert or a pre-movie advertisement at the cinema, you are likely to stumble across a car commercial. This will usually follow a predictable script that, when it doesn't dwell on the safety credentials of the car or its low energy consumption, will titillate the ego of the driver by celebrating the perspective owner's uniqueness, individuality and freedom.[12]

It has been like this for a least a century, since General Motors' supreme Sloan applied the 'model year' concept to the automobile[13] thus consecrating it as a product[14] of consumerism par excellence. The history of the car is entangled with that of consumerism at a deep level because its manufacture gave birth to the assembly line which, in turn, enabled mass production and mass consumption – a system that goes under the name 'Fordism' after Henry Ford, the most famous of the automobile industrialists.

There is another important connection between the car and consumerism, as the automobile enabled the growth of temples of consumption like the shopping mall as well as new lifestyles involving holidays away from home and leisure trips. The marriage of car and consumerism was ostentatiously celebrated in the years of the economic boom that followed World War II. Since then, the automobile has become a ubiquitous presence and has been sanctified as a symbol of success and independence. The image of a car parked outside the garage of a detached house in leafy suburbia is a classic of the baby boomers' iconography. The car also became a symbol of emancipation, romantic experience and rebellion. The resulting cultural and sub-cultural references on the topic range from Jack Kerouac to James Dean, from Crash to

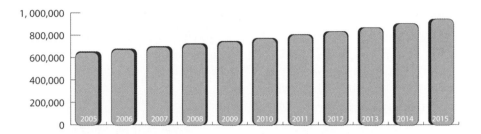

FIGURE 1.3
THOUSANDS OF PASSENGER CARS IN
USE. SOURCE: OICA (INTERNATIONAL
ORGANIZATION OF MOTOR VEHICLE
MANUFACTURERS)

Grand Theft Auto. The social, cultural, economic, physical and environmental impacts of cars have been dissected for decades by writers, journalists and academics, and interest in the subject reflects how ingrained the car is in our lives, to the extent that more than 130 years after Karl Benz patented the first automobile many people maintain the existence of a 'right to drive' as if it were a fundamental of freedom, self-determination and expression.

This lengthy love affair with the car is not due to end in the foreseeable future. Recent years have witnessed interesting changes, especially the fading interest of the younger generations in driving, so that getting a licence is no longer felt to be a rite of passage.[15] We are also seeing disruption to the car manufacturing industry. In pure consultants' speak KPMG captures this state of flux: 'traditional car manufacturers are lost in translations between evolutionary, revolutionary and disruptive key trends that all need to be managed at the same time'.[16] These changes relate to cars themselves, which are heading towards electrification, sharing, connectivity, yearly updates, and ultimately autonomy. They also relate to the

sophistications of younger and more technically savvy users; competition from new players entering the market; the growing importance of non-Western markets; and a continued increase in personal mileage.[17] However, these trends correlate with the way vehicles are used and owned and fail to question the existence of cars. In fact, current trends suggest the opposite: private car transport accounts for three-quarters of all passenger mobility. The most recent forecasts on transportation at global level are also striking. According to the *Global Mobility Report,*[18] by 2030 annual passenger traffic will exceed 80 trillion passenger-kilometres (a 50 per cent increase) while global freight volumes will grow by 70 per cent and an additional 1.2 billion cars will be on the road by 2050 – doubling today's totals.[19]

Car space

As cars continue to be deeply embedded in modern society, it is a challenging task to explore their impact on the city in all its nuances, even if the focus is restricted to one aspect such as the physical environment. Under this lens four primary areas of the car's influence on the city can be identified: *dimension, structure, form,* and *quality.*

Dimension

*If more cars are inevitable, must
there not be roads for them to
run on? If so, they must be built
somewhere, and built in accordance
with modern design. Where? This is a
motor age, and the motorcar spells
mobility. Is the present distinction
between parkways, landscaped
limited-access expressways,
boulevards, ordinary highways,
and city streets unscientific? If so,
what do the critics propose as
a substitute?*
Robert Moses, 1962 [20]

If in 2015 we had parked all the
passenger cars in circulation in a single
line, one after the other,[21] the line would
have covered approximately 11 times
the distance between the Moon and the
Earth, or 106 times the circumference of
the Earth. This is an enormous amount
of space, but it is clearly only the tip
of the iceberg if we take into account
the space required for circulation and
what we call 'redundancy', since space
is also needed at the two ends of each
car journey to park the car as well as
to ensure extra capacity in case of
peaks in demand. It is an extremely
space-hungry system, inefficient and
expensive to maintain. The issue is
magnified by the actual use of the car,
which in fact remains parked for 95 per
cent of the time.[22] Therefore, the first
obvious yet essential spatial relation
between the car and the city is one
of *space occupation*, which generally
reduces urban densities, except where
it is commercially viable to deliver car
parking in multi-storey buildings or
large basements. The result is evident
in almost all masterplans, where
car space (the carriageways and the

parking spaces) easily takes up 25 per
cent of total site areas and even more
in suburban contexts where minimum
parking requirements are fixed by local
policies and land is cheaper.

The second relationship between the
car and the city is *sprawl*.[23] In Suburban
Nation, Andre Duany[24] identifies six
main characteristics of sprawl:

1 Development organised in
   homogeneous low-density residential
   areas and a series of monofunctional
   pods (typically shopping centres,
   office parks, and residential clusters).

2 Total reliance on car and the
   subsequent creation of a pedestrian
   unfriendly environment.

3 Limited permeability within each area
   with abundant cul-de-sacs resulting
   in congestion on local distributor
   roads.

4 Generous provision of car
   infrastructure and ubiquitous
   presence of parking lots.

5 Loss of relationship between building
   frontage and the street with buildings
   pushed back from the street and
   orientated away from it, resulting in a
   loss of character and overlooking.

6 Scattered and ill-defined provision of
   open spaces often used as buffers.

These new amorphous urban patterns
were also at odds with the idea
of mixed-use centres and eroded
the relevance of traditional places
of socialisation. The spatial and
socio-economic side effects of this
phenomenon were already clear in the
first half of the 20th century (see Figure
1.4), and the concept of creating green
belts around major urban areas was a
direct response to growing concerns
about sprawl.

- - - - - - -   Dedicated pedestrian paths

FIGURE 1.4
NEIGHBOURHOOD LAYOUT
DESIGNED IN THE LATE
1920S FOR RADBURN,
NEW JERSEY – A 'TOWN
FOR THE MOTOR AGE'. A
GLORIFICATION OF THE
CUL-DE-SAC FEATURING A
FULL SEPARATION BETWEEN
PEDESTRIAN AND CAR
MOVEMENT

In the UK the idea of using green belts to prevent sprawl, maintain separate identities for urban centres and provide recreational space for city dwellers was conceived in the 1930s and implemented after World War II when power was given to local authorities to prevent development on the land surrounding cities. In London the green belt featured in Abercrombie's 1944 Greater London Plan. An additional justification for the introduction of green belts was made by urbanist Raymond Unwin who wanted to avoid creeping ribbon developments along road corridors, which could make access to the land at the back of proprieties impossible or suboptimal, and so affect their potential development.

Even if suburbia was not technically invented by cars, cars made possible the redistribution of land uses which were no longer constrained by the short range of walking. Thus cars became an essential tool of zoning and ultimately accelerated and amplified sprawl. Importantly, sprawl and zoning triggered a vicious cycle of car-dependency, which is probably the most difficult legacy of the 20th-century city. The car made it possible to spread city functions over great distances and to locate homes away from workplaces, and by doing so created an environment fully reliant on private mobility. The low densities associated with sprawl completed the lock-in effect for vast swathes of cities as the provision of public transport became excessively expensive for small catchment areas.

Structure

To approach a city, or even a city neighbourhood, as if it were a larger architectural problem, capable of being given order by converting it into a disciplined work of art, is to make the mistake of attempting to substitute art for life. The results of such profound confusion between art and life are neither life nor art. They are taxidermy.
Jane Jacobs, 1961 [25]

Enabling the car in the urban environment required the construction of a dedicated new infrastructure that included bypasses, ring roads, flyovers and urban highways. These thoroughfares profoundly changed the structure of the city – not just the urban armature but ultimately the way the city was organised and functioned. A powerful line of thought stretching from Le Corbusier to Colin Buchanan even believed that the role of urban highways was not limited to transport (see Figure 1.5). In the words of Peter Hall: *'in his London plans of 1943 and 1944, Abercrombie had sought to use new urban highways not merely to alleviate congestion, but to help define the identity of the neighbourhoods of the giant metropolis'.*[26]

Planners then started to shape large areas of suburbia (and new towns) around a strictly hierarchical system of roads, typically articulated as primary, district, and local distributors as well as access roads designed to avoid through-traffic in residential precincts. According to this logic, district distributors would also define the boundaries of different neighbourhoods and were often subject to special environmental treatments.[27]

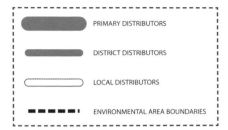

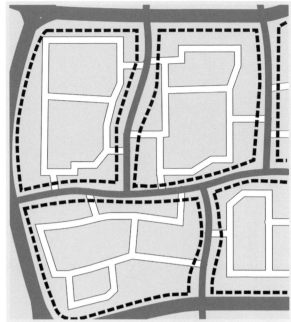

FIGURE 1.5
BUCHANAN'S EXPLANATION
OF MODERNIST PRINCIPLES
OF ROAD HIERARCHY
AND NEIGHBOURHOOD
DEFINITION. (REPRODUCTION
BASED ON THE ORIGINAL
BUCHANAN IMAGE)

Transport function came to determine street status, sweeping away the street's symbolic, economic and social roles and beginning to govern decisions on street design, starting with width.[28] Streets also had to facilitate the movement of vehicles at higher speeds, requiring larger radii and making closely-knit urban blocks unsuitable for cars. This led to profound changes in the fabric of many old city centres on an unprecedented scale, even compared to the radical effects of the introduction of railways in the 19th century.

In the years of Modernism to the constraints posed by the existing urban fabric led architects to look for solutions in radically new urban forms, including highly intricate multi-level design that separated cars from pedestrians, and many architects and planners, given the chance, advocated wholesale redevelopment.

In fact the dominant approach to city planning and design in the last quarter of the 20th century delivered only a fraction of the grand visions recommending the wholesale redevelopment of city centres, not least due to the costs involved. However, this thinking still impacted on the structure of urban areas around the world and, most importantly, still influences the work of many architects and planners. Despite the enthusiasm of politicians, developers, planners and architects, proposed redevelopments were not always welcomed by the local communities being 'regenerated'. The residents and businesses of many cities around the world fought battles against urban megaprojects, which were often triggered by the creation of new highways. Classic successful examples of public opposition are Covent Garden in London and Lower Manhattan in New York (see Figure 1.6).

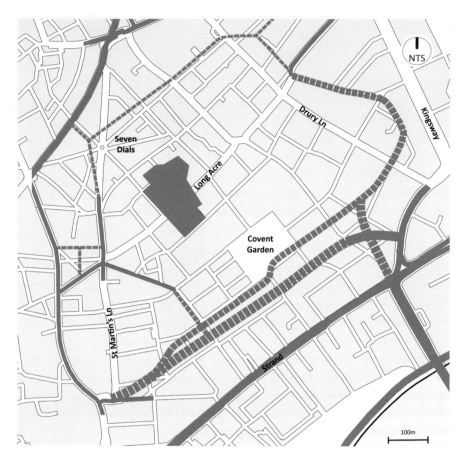

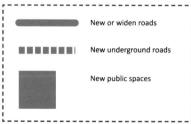

New or widen roads

New underground roads

New public spaces

FIGURE 1.6
EXAMPLES OF FAILED
MODERNIST ATTEMPTS TO
RECONFIGURE CENTRAL
DISTRICTS OF LONDON (COVENT
GARDEN) AND NEW YORK CITY
(LOWER MANHATTAN) IN THE
1960S AND 70S

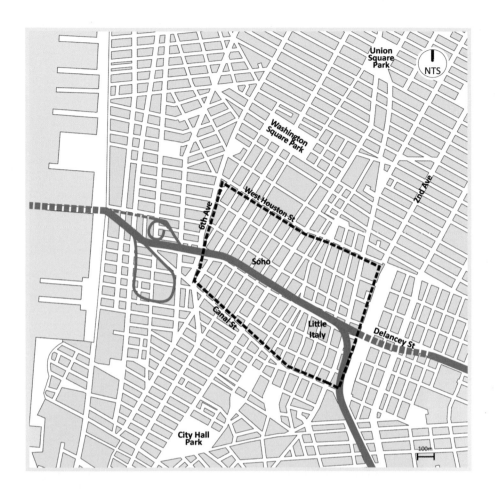

SEDDON
HIGHWALK EC2

CITY OF LONDON

FIGURE 1.7
ELEVATED PEDESTRIAN PATH AT THE BARBICAN, LONDON. THE BRUTALIST ESTATE WAS BUILT
BETWEEN THE 1960S AND THE EARLY 80S AND FEATURES A SERIES OF ELEVATED WALKWAYS TO KEEP
PEDESTRIANS SEPARATED FROM VEHICULAR TRAFFIC. THE BARBICAN WALKWAYS WERE MEANT TO
INTEGRATE INTO A 30-MILE NETWORK PLANNED ACROSS THE CITY OF LONDON BUT ONLY A SMALL
PART WAS DELIVERED.

Form

*The right to access every building
in the city by private motorcar,
in an age when everyone owns
such a vehicle, is actually the
right to destroy the city.*
Lewis Mumford, 1961

The car also redefined long-established aesthetic criteria in urban design. In the context of a functionalist street hierarchy decoupled from socio-economic considerations, the increased focus on cars not only resulted in more generous street widths and new junction geometries, it also eroded the relationship between the street and the building.

A continuous building line was once regarded as a basic design principle for defining streets and public spaces. At the turn of the 20th century the dissipation of street walls was standard practice, while in Garden City contexts the building line was separated from the street line, setting the precedent for later Modernist dogma. The spread of the automobile accelerated the erosion of the relationship between the street and its buildings and completed the obsolescence process of the traditional principles of composition: buildings were not only set back but often orientated with main entrances away from the street. Pedestrian routes along the street were also severed to allow cars to access the plots.

The idea of street design as an art was prevalent amongst influential planners at the beginning of the 20th century, notably with Raymond Unwin who believed aesthetic principles had to be reconciled with the functional needs of modern road engineering in terms of traffic and sanitation. In this sense,

Unwin's conception of the street has been defined as 'modern, but not Modernist'. It was also spectacularly different from that of Le Corbusier whose tendency was to open views and ignore traditional recipes for good street design.

*Suppose we are entering the
city by way of the Great Park
[...] Our fast car takes the special
elevated motor track between
the majestic skyscrapers: as we
approach nearer, there is seen the
repetition against the sky of the
twenty-four skyscrapers; to our left
and right on the outskirts of each
particular area are the municipal
and administrative buildings;
and enclosing the space are the
museums and university buildings.
The whole city is a Park.*
Le Corbusier, The Radiant City, 1933

The street-less city that separates users on distinct levels is a truly Modernist concept. However, as observed by Spiro Kostof, Le Corbusier's approach was only the natural evolution of an established line of thought inspired by the engineering marvels of the second half of the 19th century, which created multi-level solutions to host metro and utility networks in world capitals including Paris, New York and London.[29] Wholesale urban redevelopment and the creation of multi-level cities that separated uses as well as cars and pedestrians were seen as wholly beneficial. Buchanan, for instance, believed that the design process required a whole new outlook, 'a new synthesis of professional skills, for it is neither designing road nor designing buildings, but designing the two together as a unified process'. He called this 'traffic architecture'.[30]

By the middle of the 20th century the automobile had evolved from a simple means of transportation into a habitat for humans, and then into the true inhabitant of the city. The car required a whole new range of services and buildings including multi-storey car parks, service stations and out-of-town shopping centres.[31] Even established building forms could not avoid the effect of the car, such as the terraced house whose front garden was repurposed into parking space. The existence of driveways and space to park cars in front of buildings became boons, adding value to properties, whether their uses were residential, office or retail. The automobile revolution also reshaped the way urban corridors were organised and designed, as their architecture responded to the gaze of drivers and adapted in scale and form to the speed of cars as well as their passengers' perceptions of the environment.[32] New and redefined building forms and other car facilities began to define a city's character in new ways.

Quality

It's the sense of touch.

What?

Any real city, you walk, you know? You brush past people. People bump into you. In LA, nobody touches you. We're always behind this metal and glass. I think we miss that touch so much that we crash into each other just so we can feel something.

Opening scene of the movie Crash, 2004

Urban design, planning and architecture all shifted from catering for pedestrians to enabling the car, perversely inverting the hierarchy of functions between buildings. The dominance of traffic in urban streets has led to a deterioration of environmental quality for pedestrians in terms of noise, pollution, accessibility and, equally importantly, loss of aesthetic value. In light of this maybe it is not so surprising that the concept of jaywalking emerged in the US. Originally 'jaydrivers' were those driving on the wrong side of the road, but the term disappeared to re-emerge as 'jaywalker' referring to the pedestrian practice of crossing roads at non-designated crossing points, which is now illegal in many countries.[33]

However, the idea of resolving the issue by separating vehicles and pedestrians sucked the life out of many pedestrian environments, killing what Jacobs famously called 'the sidewalk ballet'. The attention to the quality of the pedestrian environment has undoubtedly improved since the 1960s but the 'separation philosophy' still re-emerges periodically, although generally in the context of private schemes where time and resources are available to curate spaces.

Another often overlooked implication of the automobile on the quality of the urban environment is the barrier effect on social relationships. In the 1969 seminal work The Environmental Quality of City Streets: The Residents' Viewpoint, published by the *Journal of the American Planning Association*, Appleyard and Lintell demonstrated the inverse relationship between the volume of traffic and the number of acquaintances between residents of a street – traffic creates urban fractures as much as the infrastructures that allows it, and not just physically.

The automobile also undermines 'the cohesive social structure of the city'[34] by amplifying private space in a quintessentially public space, the street. As pointed out by John Urry,[35] cars are a means of habitation: 'the environment beyond the windscreen is an alien other, to be kept at bay through the diverse privatising technologies which have been incorporated within the contemporary car. These technologies ensure a consistent temperature, large supplies of information, a relatively protected environment, high quality sounds and sophisticated systems of monitoring which enable the car driver to negotiate conditions of intense riskiness on especially high-speed road'. Cars are islands and the experience of the surrounding environment becomes mediated – the passenger becomes a spectator who does not take part in or contribute to its life.

The automobile also demands that drivers are able to read their surrounding urban environment and in this sense highly standardised road infrastructure also tends to corrode local character and has a powerful effect of homogenisation. It is not coincidental that, at the end of the century of the car, anthropologist Marc Augé uses as examples of 'non-place' much of today's car-related infrastructure, including superstores, motorways and gas stations and multi-storey car parks and contrasts them with 'anthropological places', which are historical and concerned with identity.

CAR RESISTANCE

If you want to change the world, you can start by building a bike lane.
Janette Sadik-Khan, 2018

The motorisation of cities and lifestyles that unfolded quickly and relentlessly over the last century ended up transforming most streets in urban environments into roads – utilitarian ways to serve fast vehicular movement and connect point A to point B. It left behind, in the process, the fundamental interdependence of the street with the built form, human scale, and related social life, and ultimately undermined the walkability of public urban spaces. Rapid modernisation imposed demands that traditional pedestrian patterns could not accommodate. In existing urban cores, new auto-centric urban planning principles overlaid organised grids onto organic urban fabrics and segregated formerly integrated uses. The result is that in many cities walking has become a marginal and discouraged activity. What was once a central organising factor of city life is increasingly confined to recreational or luxury activity. This transition has occurred extremely quickly in relation to the evolution of urbanity, and almost everywhere it was and still is regarded with indifference. Both in the past and present centuries, cities around the globe have been constantly redesigned for cars. The outcome is, in most cases, a rational yet generic and soulless built environment and a loss of spatial identity and sense of place.

A resistance to car-culture excesses started to emerge in developed countries from the 1960s. The influential 1963 British government report *Traffic in Towns*, the result of a study led by Colin Buchanan, shaped most of the UK's transportation planning for years beyond. The report was among the first to promote a greater emphasis on the protection of town centres and neighbourhoods from the excesses of vehicular traffic. Starting from the 1970s, in an international political climate directed toward liberalisation, deregulation and privatisation (that still persists today), increasing public and political resistance emerged to the emphasis on policies of roadway expansion and car-driven urbanism. In 1972, in *Towns against Traffic*, Stephen Plowden offered one of the first comprehensive critiques of the conventional traffic planning approach of 'predict and provide'. He was one of the first to argue that increasing road supply simply increased demand for driving.[36]

These reactions started shifting the focus of policy towards improving the overall quality of urban life, instead of focusing only on traffic-flow issues, and promoted traffic calming and pedestrian zones in urban areas as well as the use of the bicycle as a sensible alternative to cars. Traffic calming movements started in Europe and spread across the globe in the 1980s and 90s. Examples of resistance to motorisation, with experiences toward new paradigms and more sustainable transportation systems, have emerged since then around the world across scales and geographies and are widely documented. Active mobility options like walking or cycling in synergy with an efficient mass transit system are increasingly promoted as a viable alternative to the car, as they offer net positive impacts on society and help reduce health costs and produce virtually no pollution.

In addition a counter-reaction and a 'place-making' urban design philosophy took shape in response to the drastic car culture-driven transformations of the 1950s and 1960s. In the US we can trace its roots in the ground-breaking

work on the interdependence of urban design and social life by the urbanists Jane Jacobs and William H. Whyte, more recently carried forward by the Danish urbanist Jan Gehl. Jacobs advocated citizen ownership of streets through the idea of 'eyes on the street' while Whyte expounded the elements essential for creating social life in public spaces. The term 'place-making' came into use in the 1970s among professionals dealing with city making, and it assumed even stronger ideological connotations within the contextualism and postmodernism theoretical backgrounds of the 1970s and 1980s.

In North America, Vancouver in British Columbia and Portland in Oregon, after more than 30 years of comprehensive planning, emerged from being typical American auto cities to becoming models for alternative and sustainable transportation and for Transit Oriented Development (TOD).

The city of Curitiba in Brazil, and more recently Bogotá in Colombia, showed the way toward alternative and sustainable mobility systems, and highlighted the crucial integration between land use patterns and public infrastructures. In Asia in recent years, Seoul in South Korea – a mega city of 10.4 million inhabitants – has made impressive advancements toward car-resistance policies and in developing a sustainable transportation system. The Cheonggyecheon River restoration project, for instance, with its recreation of the old river bed as a linear urban park and the demolition of a 5.8 km elevated four-lane freeway, is probably the boldest example of 'trip de-generation' ever undertaken and is only the flagship project of a much wider mobility paradigm for the city. In Europe probably the world's most famous successful example of car-resistance is Copenhagen in Denmark. Strongly inspired by the ideas of urbanist Jan Gehl, the city is aiming to become the 'world's best city for cyclists'. Notable also are the recent successes in sustainable transportation in London and Paris and at different scales of Freiburg, Karlsruhe and Stuttgart (German's car capital), to cite just a few.

With the rise of environmental and global warming concerns, from the publication of Rachel Carson's *Silent Spring* in 1962 to the Kyoto Protocols of 1997 up to today, resistance towards car dominance has gained an even more multi-layered significance as the German city exemplars illustrate well. Germany has in fact proved that the modern combustion engine automobile must die. A recent article from *The New Republic Magazine*[37] illustrated the role and impact of the current automobile system on development patterns and climate change. Germany, as a case in point, is today at the forefront of larger developed countries in the fight against global warming in accordance with the Kyoto Climate Change Protocol. So far, Germany has reduced its greenhouse gas emissions by 27.7 percent—an astonishing achievement for a developed country with a highly-developed manufacturing sector. Despite the country's impressive efforts to reduce greenhouse gas emissions, it is short of its ambitious goal of reducing emissions by 40 per cent by the year 2020. The reason for this may well come down to cars. In 2010 a NASA study declared that automobiles were officially the largest net contributor of climate change pollution in the world. Transportation in the US, where more than 90 per cent of daily trips are by automobile, is now the largest source of carbon dioxide emissions, according to an analysis from the Rhodium Group. In Germany, the transportation sector remains the nation's second largest source of greenhouse gas emissions, and may soon become the first. Germany has an established car culture and, after the US and Japan, is the largest car maker in the world. In order for Germany – a multimodal haven compared to the US – to meet its ambitious emissions targets, it will require a massive lifestyle change that will entail half of the current driving population to switch to other, more sustainable modes. In other words it will require the adoption of a lifestyle and *a resulting urban form that is not car-ownership dependent*, and drastic public policies will be required. High-emission cars in densely populated areas will have to be banned while significant investment in alternative public transportation infrastructure will be required. The switch to an electric CAV car system is an alternative solution envisioned by many car makers today, and one that may require relatively modest infrastructure investments. The crucial question is how electricity will be produced, as only a switch to renewable sources will effectively contribute to a reduction in carbon emissions. It will, however, be an important step towards cleaning the air of the world's urban environments in a context where it is estimated that around 53,000 US citizens die prematurely from vehicle pollution each year.[38]

THE RISE OF AUTOMATION

The street is the river of life for the city. We come here not to escape them, but to partake them.
W. H. Whyte, 1979 [39]

The rise of the driverless car is likely to follow a similar trajectory to that of the introduction of the automobile – it will be a bumpy road that becomes smoother with time. Public acceptance may in fact not be as great of a challenge as it might appear today, despite setbacks such as the first recorded pedestrian fatality involving an autonomous car, which took place in Tempe, Arizona in March 2018, claiming the life of Elaine Herzberg.[40] Historically, the safety of cars has been an issue of contention, with those opposing cars decrying their danger despite the surprisingly high number of fatalities linked to horses.[41] The safety of cars improved as they became more affordable and more irresistible and the rest (including the often opaque behaviour of large car manufacturing companies) is history.

There are a number of other similarities between the emergence of cars and CAVs, including the role played by military technology, vehicle challenges, marketing and corporate communication. However, we should not think of CAVs as an absolute novelty. In fact CAVs are only slightly younger than cars, with people first imagining a future of completely automated cars shortly after the invention of the car. In a way, extending the 'auto' from the vehicle to the act of driving itself is a logical step, despite the technological challenges.

It is beyond the aim of this book to look in depth at the story of CAVs,[42] however there are aspects worth noting as they provide insight into the relationship between the emerging form of mobility and cities. Today, the development of driverless cars seems almost like an underground story running parallel to that of the automobile. At the 1939 World's Fair, General Motors presented Futurama,[43] a vision of a 1960s city crossed by automated highways populated by cars guided by radio control systems. The technology may not have been there yet, but the vision proved a popular one. During the course of the 1950s General Motors tried to pursue this vision, pioneering experimentations with electric highways that continued with trials in the 1960s and the 70s. These efforts were expensive and limited by the technology of the time, so they never managed to advance the driverless car beyond the concept stage. It was in the final decades of the 20th century that the modern driverless car began to emerge from the labs of robotics researchers.

Throughout the 1980s and 1990s, German autonomous-vehicle pioneer Ernst Dickmanns built several prototypes that used sensors and intelligent software to steer themselves. In Italy, Professor Alberto Broggi created a car that used machine vision software to follow painted lane markers. As primitive as these early driverless cars were by today's standards, they had a major advantage over the radio-guided Buicks of the past: their intelligence was carried onboard the car rather than buried in the road.[44]

Two catalysts helped to make modern robotic cars a reality: the exponential growth of the power of microprocessors (and their miniaturisation); and US military targets to automate vehicles in war zones. In the previous two decades, a number of technologies such as advanced sensors were perfected while others such as artificial intelligence and deep learning evolved dramatically. This convergence created the conditions for the driverless car as we understand it today.[45]

In the second half of the 20th century focus shifted away from the creation of a centrally controlled environment able to determine the behaviour of 'dumb' vehicles. Current research and development is concentrating intelligence on vehicles that aspire to read the environment and any unexpected conditions that may materialise. The implications are substantial – not just for the operation and technology of the driverless car – but also for the environment in that it would no longer need to be cabled and retrofitted with a complex layering of sensors and communication devices. From this perspective, the urban environment will be free to evolve independently from an expensive infrastructure that could also become obsolete very rapidly. The relationship of vehicles and the environment could also be enhanced by enabling the exchange of information – with pedestrians, cyclists, traffic lights and sections of roads in bad repair, for instance, all sending alerts to CAVs. Overall the modern interpretation of driverless cars promises to take most of the complexity on board, leaving the streets themselves relatively 'light' in technology. At least from a technological perspective, it appears that cities will not need heavy retrofitting to accommodate CAVs.

Great coordination effort will however still be needed to manage a network of streets used by CAVs, or even more so by a mixed condition of CAVs and cars, potentially for a long transition phase. This will trigger a whole series of critical decisions regarding who should or could run and be accountable for such a system.

CAVs could potentially contribute to an acceleration towards smart cities[46] that could also have important spatial side effects. Adam Greenfield explains in *Against the Smart City*[47] that the problem is allowing technologists to design the urban environments as what Deleuze called 'any-space-whatever' – that is abstract, generic, unconditioned spaces, containing infinite potentials for connection. The common image of the smart city is built on a language that essentially replicates most if not all of the blunders we associate with the discredited high-Modernist urban planning techniques of the 20th century. This could conjure another risk, that of attempting to reduce complexity by creating a predictable environment where CAVs can operate – a dedicated infrastructure that segregates the CAV from other forms of transport. This is a potentially disastrous scenario, evocative of the 1960s visions just discussed. The evolutionary trajectory of the CAV, and the way it collides with other technological and socio-economic trends is important in influencing the way the CAV relates to the urban environment and defines some of its future characteristics.

Chapter 2

SETTING THE SCENE

Driverless cars are a fantasy as old as the car itself. However, the technologies required to realise the vision only very recently converged and came to fruition. Such technologies are rapidly evolving and improving, requiring a continuous learning process from all stakeholders. In their current form, the technologies still present some reliability issues – particularly under severe weather conditions – although it is expected that these will be overcome in the near future. The American Planning Association in a 2018 report defined autonomous vehicle technology as:

An umbrella term that includes a wide variety of features and technologies that enable vehicles to take control of some or all of the major driving functions normally completed by the driver. This includes fully autonomous vehicles that no longer require a human driver to operate them, as well as a range of advanced driver assistance systems (ADAS) that enhance driver safety by taking temporary control of one or more driving functions (speed, lane position, braking, etc.). A fully autonomous vehicle no longer requires a human operator to drive.[1]

THE AUTONOMOUS VEHICLE

Autonomous vehicle technologies in fact comprise a number of sensors, cameras, light detection and radar – Lidar or light radar – and advanced software that combines remote sensing, recognition algorithms, network analysis, and previous data 'experience' drawn from other CAVs. Cameras and other sensors provide supplemental detection systems, particularly for the area immediately surrounding the vehicle. More sophisticated systems can also assess and predict where vehicles and pedestrians will go next, for instance starting to slow down before a pedestrian even enters the street.[2] Waymo LLC – a self-driving technology development company that originated as a Google project and is now a leader in the advancement of driverless cars in North America – defines fully self-driving vehicles as a combination of various technologies that enable the answers to four fundamental questions: Where am I? What is around me? What will happen next? And what should I do?

Starting in April 2017, Waymo initiated a limited trial of a self-driving taxi service in Phoenix, Arizona. Around 400 early adopters used the service in its first year, increasingly with no human safety driver. Waymo's vehicles are equipped with detailed three-dimensional maps that highlight information such as road profiles, curbs and sidewalks. Sensors and software scan constantly for objects around the vehicle such as pedestrians, cyclists, vehicles, road works, obstructions and also continuously read traffic controls. The vehicle's software predicts the movements of everything around it based on their speed and trajectory, using that information to predict the many possible paths that

other road users may take. Based on all this information, the software determines the exact trajectory, speed, lane, and steering manoeuvres needed to progress along the route safely.[3]

Connected Autonomous Vehicles (CAVs) are vehicles that integrate both autonomous and connected technologies. It is the combination of these technologies that promises to bring the most radical and potentially positive changes to cities. Connected vehicles are equipped with technology enabling them to connect to devices within the car, as well as to external networks such as the internet, or to advanced traffic management systems (ATMS).[4] This technology relies on information gathered by both vehicles and the transportation infrastructure about real-time operations of the transportation network. In theory CAVs can avoid traffic jams altogether by finding the fastest alternative route in real time. They can also inform other vehicles of their intended movements, travelling in a safer and more harmonious way than today's traffic. This will enable more efficient management of the existing road network and of the fleet of vehicles in cities.

The continuous rise and improvement of automated vehicle technologies is partially due to their potential benefits to the environment and to humans, such as improved safety, increased accessibility and better air quality (if CAVs are electric). These factors may help to accelerate adoption by overcoming cultural obstacles. Another important driver is the enormous amount of money vested in disruption of the current model of the automobile industry. There has been

ANATOMY OF AN ELECTRIC CAV

FIGURE 2.1
ANATOMY OF A CAV

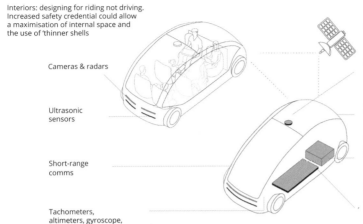

Interiors: designing for riding not driving. Increased safety credential could allow a maximisation of internal space and the use of 'thinner shells'

Cameras & radars

Ultrasonic sensors

Short-range comms

Tachometers, altimeters, gyroscope, odometry sensors

Cloud and GPS data

Lidar

Antenna for GPS and wireless data exchange

Computer processors with redundant systems for maximum safety

Electric batteries and motors simplify mechanical and structural requirements. Charging options (traditional, induction etc) also open possibility in terms of reconfiguration of the vehicle

CAVS DEPEND ON A NUMBER OF TECHNOLOGIES TO FUNCTION:

A Global Positioning System (**GPS**) provides information on positioning and plays a key role in determining the vehicle's itinerary. Its accuracy, however, is only within metres thus requiring the use of **tachometers, altimeters, odometers and gyroscopes** to perfect the vehicle positioning.

Ultrasonic sensors detect the position of objects in close proximity to the vehicle (as is currently the case for assisted parking).

Video **cameras** (including infrared for low-light conditions) aid the car's computer to build a picture of the surrounding environment from different angles. They also integrate the high-resolution 3D scan built by Light Detection and Ranging (**Lidar**) based on laser technology and effective up to 200m. For closer distances these are assisted by **Radar** sensors, which identify objects, their position, size and speed at shorter distances (the same technology used today for adaptive cruise control).
Short-range communication devices

enable vehicle-to-vehicle (**V2V**) and vehicle-to-infrastructure (**V2I**) exchange of information on traffic, proximity of junctions, and vehicle trajectories, for instance. They also update maps, update software and fix bugs, as well as updating the algorithms managing detection, decision making and the general operation of the vehicles.

All data collected by the sensors ('sensor fusion' is the integration of this information to build a comprehensive understanding of the vehicle's surroundings) is fed to **computer processors**. The processors and key technological equipment have backups to maximise safety, similar to those of planes.

SUPERCARS AND CAVS

Today's CAVs are presented as supercars, that is as traditional cars equipped with technology enabling them move autonomously.

The new technologies used in CAVs allow a rethink of the form of the car and introduce new constraints at the same time.

However, it is wrong to assume that technology alone will dictate the shape of CAVs. Future vehicles will have to integrate sensors to maximise their performance, but other factors will be relevant to design including:

- **Safety** standards if CAVs dramatically improve safety, passenger cages will become thinner and lighter

- **Ownership** or additional **uses** if CAVs are operated as part of ride-hailing fleets or used to deliver parcels, vehicles may increase in size)

- **Type of service** whether the CAV is a basic product offering just transportation or a luxury service integrating onboard entertainment, mobile office space, and so on will also determine the size and shape of vehicles.

Other currently unknown elements will also need to be taken into consideration in the design of CAVs.

As the form of the vehicles and the way they interact with their surroundings change, so will the streets that will host them.

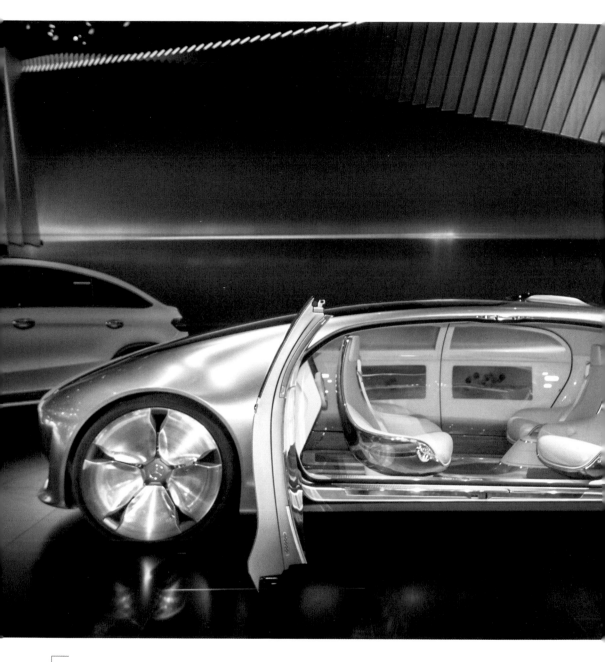

FIGURE 2.2
MERCEDES AUTONOMOUS CONCEPT CAR AT
THE NORTH AMERICAN INTERNATIONAL AUTO
SHOW, DETROIT 2015

huge investment from start-ups, IT powerhouses, and other non-traditional players as well as the traditional car manufacturers. Recently, Navigant Research released a report on the most influential autonomous car manufacturers, tech giants and start-ups racing to deliver full autonomous driving technology. The majority of the leading companies are large car manufacturers with the help of start-ups building software and hardware. Among these are Google/Waymo, GM/Cruise Automation, Uber, Delphi/nuTonomy, Lyft, Ford, Audi, Tesla Toyota, Volvo, BMW, Mercedes-Benz (Daimler), Honda, Renault-Nissan Alliance, PSA, Porsche, Hyundai, Luminar/Austin Russel and Comma.ai/George Hotz. Top-tier suppliers include Mobileye, Velodyne, Qualcomm, Intel, Nvidia, Samsung, Baidu and Apple.[5]

Many of these players have aggressive timelines for when they anticipate offering highly automated vehicles, but the adoption of AV technologies will vary significantly based on the level of automation. Level 1 features, such as cruise control, have already been on the market for a few years, while many Level 2 features are today available in luxury cars such as Tesla's models. Tesla's autopilot feature, introduced in 2015, controls the vehicle's speed and lane position. It is classified as Level 2 automation, requiring the driver to be in control of the car at all times. Level 2 automation features are likely to become widely accepted in the coming years. Nevertheless, only Level 4 and 5 automation is expected to have significant impact on the form or function of the transportation system, and therefore a significant impact on social behaviour and urban form.

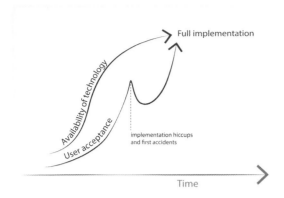

Full implementation

Availability of technology

User acceptance

implementation hiccups
and first accidents

Time

FIGURE 2.3
USER ACCEPTANCE AND
OTHER FACTORS NORMALLY
DELAY THE DEPLOYMENT OF
INNOVATION DESPITE THE
AVAILABILITY OF TECHNOLOGY

FIGURE 2.4
AUTONOMOUS BUS TESTING IN
STOCKHOLM, 2018

The latest forecasts suggest automation Level 4 vehicles will be available for purchase within three years in the wealthiest markets. This level of automation will allow AVs to perform all safety-critical driving functions as well as monitoring surrounding conditions, but it does not cover every driving scenario. So Level 4 AVs will be able to safely pull over and stop a journey if the driver fails to intervene in a situation the AV's operating system cannot handle. In such a vehicle, the driver could devote part of the time to sleeping, eating, or working on a laptop instead of driving. Level 4 vehicles are expected to reach the market around 2022 and Level 5, with full automation in all conditions, by 2030.[6,7,8] This timeline is, however, reliant on the necessary technologies and will also depend on the capacity of markets to address critical challenges in terms of the regulatory framework including insurance, as well as issues of public perception and ethics. Freight and public transit industries are likely to lead the way and would contribute to accelerating general public acceptance. The challenges are significant, not insurmountable, but they could slow down the adoption process (see Figure 2.3).

Initially, CAVs will significantly alter the cost structure of vehicular transport because of the higher costs of the technologies. It will take time to lower costs and make CAVs as relatively affordable as the automobile of today. This will result in private individual ownership being a less attractive option. On the other hand, CAVs will reduce the price of on-demand shared mobility by removing the need for a driver. This price shift in transport costs could be leveraged in favour of a significant reduction in vehicle ownership, particularly in urbanised areas where on-demand shared mobility is more feasible. Private ownership is likely to remain the dominant model in rural areas, while shared vehicles are expected to represent a significantly larger percentage of the vehicle fleet in cities. There will also be a transition period of several decades in which the old automobile and the CAV will coexist before CAVs reach full adoption. Some studies suggest that in around 15 years' time, CAVs could make up a quarter of all vehicles on the road.[9]

Meanwhile, an increasingly urbanised world is heavily dependent on the automobile to move. 2008 marked the year when 50 per cent of the global population passed from living in a rural environment to living in an urban or semi-urban one. In the next two decades, Chinese cities will absorb a new population of some 350 million people while the US will witness a population increase of over 100 million people in the next 50 years and in the UK the population will reach 73 million by the beginning of the 2040s.[10]

Viewed in conjunction with climate change challenges these forecasts clearly illustrate the need to formulate a pragmatic vision of CAVs and urban design. A transport transition could turn into a nightmare if not managed and prepared appropriately. Sprawling development, congestion, inequality, privacy issues and job losses are among the most pressing challenges that cities will face in this transition. Preparing for CAVs will mean securing the benefits they offer while mitigating against risks, including the risk of designing cities around the new technology rather than around people.

CAVS Sprawling

What will the new global middle class seek in a CAV-powered lifestyle? Could the CAV be the trigger for a new suburban revolution, much like 19th-century industrial London or 20th-century America, but even larger and more dramatic? As emerging and global powers like China diversify to include a rising knowledge-based economy, how will urban trends be affected by the CAV revolution? The current outlook shows a continuous and dramatic increase in auto dependence throughout most of the largest global economies, with 2016 proving a record-breaking year for global car sales.[11] In relation to the density of Chinese cities, for example, the car-oriented model is already becoming unsustainable as land-use patterns and the design of street networks create a vicious circle that discourages alternative modes of transportation and makes environmental conditions intolerable. As a result, malaise among urban dwellers in China is growing.[12] New models for a healthier environment where residents can escape pollution and the stress of an over-congested city will be increasingly appealing, but conversely CAVs could significantly worsen that congestion while providing the means of escape. This perverse effect of CAVs will be triggered by a potentially explosive combination of factors including a rise in the number of people using vehicles (as a driving licence won't be needed), an unanticipated increase in private ownership of CAVs, and increasingly larger urban areas. Commuting is likely to be tolerated for much longer journeys with CAVs, as travellers will be able to use journey time for other activities or for sleeping. Commuting could also be cheaper, thanks to shared CAVs systems. The increased efficiency and convenience of CAVs will encourage people to live even further away from city cores and jobs.

If the average commuter is currently unwilling to travel for much longer than an hour, a common reasoning in favour of CAVs will be along the lines of: 'my current commute consists of one hour of continuous driving for 30 miles; with a CAV it would consist of 1.5 hours of working or sleeping while the vehicle drives itself 50 miles'.

CAVs may also reduce travel cost per mile if the commuter will not have to purchase or maintain the vehicle. People will then be able to afford to live even further away from their jobs, enjoying more space while spending less. These factors will extend the catchment areas of metropolitan regions and will trigger sprawling real estate development in previously untouched rural areas, in a scattered pattern only dependent on the availability of a road network. By providing access to an alluring new 'middle landscape' of escape, CAVs could deflagrate urban cores and ramp up sprawl to unprecedented levels, undermining the recent revival of the inner city worldwide. Additionally, as successful urban cores become increasingly unaffordable, the need to access affordable housing away from jobs in city cores is likely to be exacerbated.[13]

Western suburban history provides ample warning of what can go wrong. The list is long – *suburbs breed environmental hazards, energy-intensive development patterns, over-dependence on vehicles, gated communities, urban decay, privatisation of public space, endless sprawl*, as well as *obesity and related health problems*. Yet Western history also provides successful alternative models, promoted among others in the work of the Urban Task Force in the UK or, more recently, by the New Urbanist and Smart Growth movements in the US. It is now accepted (almost everywhere) that walkable environments and mid-density characteristics make public services and public transport more economically feasible and preserve a human scale while promoting sociability and leaving room to beautifully integrate city and countryside. These are all precious lessons to keep handy in preparing for CAVs.

CAVs technology could provide opportunities for promoting more compact development patterns by leveraging recent urbanisation trends such as Transit Oriented Development (TOD) and lower rates of car ownership and reduced vehicle fleet in cities, particularly if CAVs are shared. In fact, recent studies demonstrate that one shared CAV could replace up to 10 privately-owned vehicles.[14] This will not necessarily mean a reduction in traffic congestion, however, as each CAV will be travelling significantly more miles per year then an average automobile today.

CAVS and the urbanists

Driverless cars have been on the radar of architects and urban designers for several years. However, the topic has generally been approached with a certain degree of simplicity, and the visions produced so far have typically culminated in an idyllic, greener view of future streets. These optimistic visions have an important role to play as they build a collective and desirable image of the future, thus setting expectations and influencing future ideas for development. At the same time, they also tend to focus on the public realm, leaving behind the implications for building design and also the key interaction between buildings and public space, or they ignore the strategic aspects of road networks and the different functions of streets. Furthermore, most of these visions transpose current ways of thinking about the design the public realm into the future, adapting and redistributing space rather than rethinking it. They also tend to jump directly to the long-term scenario, thinking more about the opportunities – such as the fewer parking spaces required or the improvement of air quality – rather than the issues that CAVs may create.

FIGURE 2.5
AUTONOMOUS VEHICLE
TESTING IN PARIS, 2017

There are risks associated with this somewhat narrow approach to CAVs, especially in conjunction with an equally slow engagement with city planners and politicians. Crucially, designers risk missing the opportunity to make a real impact on the debate by allowing the transition to be driven by the agendas of technologists and businesses, which potentially conflict with public interests. There is also the potential to replicate mistakes made in the Western world, especially after World War II, when our urban areas were moulded around the car. Could a similar scenario be repeated if cars are replaced by CAVs? Will a separation between modes of transportation be reintroduced to accelerate their adoption? By making CAVs ubiquitous in the city, how can it be ensured that this will not undermine bicycling and walking as first/last mile options of choice? How can CAVs and cars coexist in the same streets? Is it realistic to think that extra space on the street (if this can be secured) can be turned into good quality public space? Who is going to pay for and maintain this in a world of shrinking municipal finances, worsened by the certain loss of parking revenue? What can be done to alter CAV-promoted low-density suburbs? If built environment professionals hesitate to ask these questions and don't start drafting solutions, then they will miss the opportunity to shape the debate and ensure that cities will be designed around people rather than around the needs of the new technology.

A PLAUSIBLE SCENARIO

A much-presaged benefit of the CAVs revolution, apart from safety,[15] will be more efficient traffic operations leading to an increased carrying capacity within the existing road infrastructure. This will also mean fewer parking requirements, with the potential to reallocate road and parking space away from vehicles to other uses. With the ever-increasing levels of traffic congestion that urban environments will experience in coming years, there will be a temptation to simply use the space 'gained' for more vehicles. Such a strategy would be short-sighted. It has already been proved many times in the past that more and larger roads simply lead, rather quickly, to an increased volume of vehicles. Instead, it is necessary to explore the potential for a fundamental rethinking of our city cores and to advocate among city policy makers for a paradigm shift that breaks the cycle of motor vehicle dominance on the street. We need to champion the reallocation of space away from vehicles, to promote more productive and resilient uses. To finally break the perpetual cycle of traffic-induced demand we also need to champion human-scale environments and a people-centred approach. With the coming of CAVs, local authorities and policy makers will have the once-in-a-lifetime chance to reconsider how their cities' streets function as part of a movement network – a unique opportunity to use the 'extra space' to retrofit their cities in a more context-conscious and sustainable way. CAVs could be a sustainable answer to urban transportation, opening up a post-car paradigm and creating opportunities for resilience and climate change adaptation thanks to the changes that they could bring to the physical fabric of cities. The next 20 years are the most crucial, tackling the very practical challenges of a transition between current technologies and full automation. Ways in which CAVs are implemented at their introduction will inevitably influence their subsequent trajectory and potentially create infrastructural or cultural rigidities that could affect our cities for much longer than the years of the adoption phase. This could be in terms of separation, attitudes towards first/last mile, or in other ways that cannot yet be predicted.

A likely market predication for the near future

Given the powerful set of variables at play when determining the characteristics of CAVs – how they will be used, the speed of their adoption, and ultimately their effect on the built environment – it is difficult to outline an exact future. To describe a plausible framework, a series of assumptions has been derived from an understanding of the latest market predictions.[16] Within the next 20 years the wealthiest urban environments on the planet will all present a series of common new defining characteristics:

Vehicles
Every vehicle circulating will be partially automated and at least a quarter of the overall fleet will be fully automated. This means that every vehicle will have an automation level between 2 and 5.[17] In 20 years, all these vehicles will also be connected and will exchange information with other vehicles. They will be able to permit platooning, so that space between CAVs while moving could be minimal. The majority will be electric and will require dedicated charging infrastructure. As a result, there will be a substantial decrease in pollution emitted where an electric fleet is deployed.

Generally, the CAVs used will be less than 2 metres (6.5 feet) wide,[18] while for emergency services current standards for space and dimensions will still apply.

Streets
Congestion will still be a major issue in cities, as well as in CAV-only areas. Streets will be quieter than today, with a reduction in traffic noise due to the deployment of electric engines. Traffic lights and signage will remain but will perform differently; the focus will be more interactive with a shift to pedestrians and bicyclists. New ways to regulate possible conflicts among modes of transport will emerge. Street lighting will be experiential and smart. CAVs will require frequent designated areas for pick-up and drop-off, therefore provision of these areas will increase considerably both on-street and off-street. Reductions in roadway sections will be possible thanks to improved traffic flow, with a reduction of lanes in terms of number and width, and therefore in overall roadway widths. Dedicated charging infrastructure will also be required. Only part of these will be on-street, the rest will be in dedicated areas and stations. Induction charging facilities will be initially introduced for taxis and buses. Facial recognition technology will start to be implemented. Streets could operate two-way, enhancing the overall flexibility of the network. Intersections and particularly small intersections will be more fluid, accommodating a consistently slow traffic flow as opposed to a persistent stop-and-start movement.

Parking
Kerbside long-stay on-street parking will gradually disappear, particularly in city centres. Depending on the specific context, at least 30 per cent of the existing parking supply will become redundant. Parking spaces in urban

cores will still exist near destinations but will be fewer and will be supplemented by remote CAVs parking locations.[19] Also, car parking space dimensions will be smaller than today as CAVs will require less space to manoeuvre and will be able to self-park.

Multimodality
Users' mobility needs will increasingly be provided through subscription models like a 'Netflix' of mobility services. In most cases, these services will be regulated by the public and run by private operators. Microtransit CAVs service systems will be deployed to cover first/last mile trips connecting to mass transit corridors, with a hub and spoke model. Mass transit also will be partially, if not completely, automated. Traditional human-driven taxis will increasingly be replaced by the Mobility as a Service (MaaS) model and driverless services. CAVs and bicycles will compete in urban cores for first/last mile trips. Drones for emergency services, personal mobility, and for delivery will be introduced for core-to-core and hub-to-hub trips, particularly in large metropolises with heavy congestion.

Real estate
With the introduction of CAVs in urban cores, the immediate benefits for developers and housebuilders will be reduced requirements for the construction of on-site and off-site parking areas and the possibility of building reduced road infrastructure. Fewer parking requirements will mean more land available for development. Service areas for loading and unloading operations, depending on the level of automation, will shrink. CAVs will entail different benefits depending on the type and location of buildings. There will be opportunities for substantial densification of established urban areas. They will be a game changer for out-of-

town retail and office parks but will also make urban infill sites more palatable for development, as accessibility issues and absence of on-plot car parking will lower the viability threshold of many schemes. There will be opportunities for retrofitting existing buildings, however the benefits will be lesser for existing parking structures, as underground facilities tend to have reduced floor-to-floor heights suitable for fewer uses. As parking will be decoupled from the house, MaaS type of services will be packaged as part of the lease, with offers for bicycle hire, car clubs, or credits for the local public transport system.

A positive framework for CAVS

Some of these predictions may not come to fruition or may develop on slightly altered paths. However, collectively they provide a plausible scenario for envisioning a positive framework for CAVs to test design ideas, as elaborated in the following chapters. Furthermore, in order to be a sustainable answer to urban transportation, particularly in urban cores, and to capitalise on their potential benefits for urban form, CAVs will have to combine a fundamental set of characteristics. In particular, they will have to be:

Electric
Electric-engine vehicles will eliminate the issue of pollutant emissions, improving significantly air quality in cities and possibly helping kerb CO_2 emissions, depending on the energy source.

Connected
Vehicles equipped with technology enabling them to connect to devices, as well as external networks such as the internet and vehicle-to-vehicle (V2V)[20] will make it possible to manage the use of the existing road network and the vehicle fleets more efficiently, freeing up space for other and better uses.[21]

A service
The future of mobility will be service driven by mobility providers (public or private). Passengers will give their requirements for being transported via the cloud, and the mobility provider will provide CAVs for transport. This service will be part of an integrated range of mobility options to take users from origin to destination.

Shared
CAVs services will be able to make multiple stops and pick up multiple passengers travelling in the same direction; this will lower the cost of trips while expanding the number of passengers per vehicle. It will also attract a pool of elderly, teenaged, and disabled users, helping to free up space for other uses.[22] At the same time, privately-owned single-occupancy vehicles will have to be progressively banned from city cores.

Flexible
Through public and private partnerships, CAVs services will help make fixed mass transit services more functional, efficient and accessible, particularly by filling first and last mile gaps.

Values underpinning good urban form

In developing a positive framework for cities in the age of the driverless car, urban design should always promote a series of underlying values as fundamental principles of good urban form. This is regardless of any particular technology.

Human scale and human well-being
People-scaled design and human well-being must be at the core of any consideration and decision making regarding urban design and planning, rather than technology.

Accessibility
Enabling different options to access the areas of the city for everyone is a fundamental right and an issue of social justice and democracy.

Walkability
Walking is, and must remain at, the core of sustainable transport strategies. This also applies to cycle routes.

Strengthening the social infrastructure and the sociability of places
Humans are by nature social animals and will remain so in the future. One of the fundamental goals in urban design is to promote the provision of abundant and varied public spaces and to design to support and strengthen this key infrastructure in cities.

Maximising opportunities to improve and preserve biodiversity and resilience
In the age of climate change, global warming and unprecedented urbanisation, urban design must be a practical tool to make cities climatically more resilient and to preserve and promote ecological biodiversity.

Densification as a primary response to accommodate urban growth and to control sprawl
Wherever possible, urban design should promote an appropriate densification of already developed and underused urban areas. This is to create more compact mixed-use clusters, support transit-oriented development, and to limit greenfield development.

Reintroducing places for production in the city

When uses are compatible, places for production in cities must be protected or reintroduced. This to promote a liveable, highly mixed-use environment and to combat the current phenomena of transforming cities towards being places of consumption only.

Using data to improve city living and sustainability performances – a smarter city

Urban design must make optimal use of all the interconnected information that will be increasingly available to cities in order to help them to make better use of limited resources and to control their operations efficiently.

Celebrating local identity

As already seen for the automobile, there is a real risk that CAVs will foster further uniformity and homologation in cities. Urban design must remain contextual and always be an opportunity to reveal local identities.

These values are rooted in the CAVs principles for people centric urban design discussed in the following chapter. See appendix II for an interview with Andes Sevtsuk, Director of the City Form Lab, on CAVs, urban design and form. ▬

Chapter 3

CAVS PRINCIPLES FOR PEOPLE-CENTRIC URBAN DESIGN

Based on the positive framework described in the previous chapter, this section discusses and recommends urban design principles for managing CAVs in cities, particularly focusing on their urban cores.

The eight principles are:

1. Design streets not roads
2. Keep it legible
3. Share CAVs, share streets
4. Reallocate space
5. Phase out cars
6. Enable new architecture
7. Make it resilient
8. Win fast

DESIGN STREETS NOT ROADS

Over the last 20 to 30 years many cities around the world have made huge progress in addressing the Modernist mistakes of car-led city planning, dismantling the barriers that were created when traffic efficiency was the driving force. These barriers destroyed many streets, separating pedestrians and cars, as well as cyclists (or ignored them completely). These mistakes of the past should not be forgotten in the rush to embrace the future. Cities are for people, and streets (not roads) should be designed for everybody. For this reason, CAVs should be used in a people-centric manner, in continuing to pursue the aim of improving our cities for everyone, not prioritising the efficiency of this technology over other modes and users.

Use redundant space for people, not more CAVs

One of the much-heralded benefits of CAVs will be more efficient traffic operations leading to an increased carrying capacity of the existing road infrastructure, which will provide an opportunity to reallocate road space away from vehicles to other uses.[1] However, where high levels of traffic congestion are experienced, there could be the temptation to simply use the space 'gained' for more vehicles. Instead, gained space should be the opportunity to instigate a paradigm shift, breaking the cycle of vehicle dominance on the streets, and championing the reallocation of additional space from motorised vehicles to people and human-scale activity. Another version of traffic-induced demand should not be perpetrated.

Keep decluttering

CAVs offer the opportunity to declutter streets even further, as they should not require more paraphernalia. Traffic signals, car-related signage and road markings all make streets cluttered, creating physical obstructions and visual 'noise'. Some of these items could be removed or reconsidered in terms of design once full transition to a CAV fleet has occurred. There will be, of course, a long transitional period where vehicular traffic will constitute a mix of CAVs and manually-operated vehicles. Much of the current infrastructure will need to remain within the street environment for this time. It should not be forgotten that some infrastructure benefits pedestrians and cyclists as well as vehicles.

Differentiate, don't separate

CAVs will be able to 'read' which space they can use and which to avoid using, for instance, geofencing technology. This should minimise the future need for separation by means of road barriers as well as their enforcement by cameras or humans and it should support the removal of vertical and horizontal signage and deflections. What is needed is some form of visual delineation, clear enough for people walking and cycling to know where and when they should expect to encounter CAVs, especially in shared spaces, hop-on/drop-off (HODOs)[2] or loading/servicing bays. The delineation will mostly be for people not for CAVs. For example, this could be a simple landscaped approach with variations in materials, colours, and integrated light systems warning that a CAV is approaching.

The CAV revolution needs a bicycle revolution

Important gains have been made for cycling in many cities around the world, and it should continue to be so for environmental and health reasons.[3] However, in the short to medium term there will be a conflict between CAVs and cyclists. Interactions between CAVs and cyclists will inevitably be more complex and nuanced than those between CAVs and cars. CAVs will eventually learn how to respond to cyclists, however there is likely to be conflict in terms of requirements and at best a much-reduced level of efficiency for CAVs due to potential extreme stop/start operation in city centres. This should be asked to provide first for cyclists and then for CAVs (and cars) to avoid conflict. Catering for pedestrian and cyclist needs will mean making streets more predictable, and thus more usable by CAVs. Tough choices about space allocation will be unavoidable. Dedicated cycle lanes should become the norm on all higher order streets, with sensors that communicate to CAVs that they cannot enter the lane space. Over time cycle lanes will not have to be designed with barriers but could be simply demarcated visually in CAV-only zones. In lower order streets CAVs should be speed limited and space can be shared with cyclists with no need for separation.

Manage urban platooning

'Vehicle platooning' will be the use of radar and vehicle-to-vehicle communication to form and maintain a close headway formation between two or more in-lane CAVs. Platooning could permit a more efficient use of road space as well as saving energy while increasing road capacity.[4] In urban cores, CAVs should be instructed to avoid the 'wall effect' of endless in-lane vehicle formations, similar to what is experienced today with automobiles during rush hour but at lower speeds. Platooning could be extremely disruptive to other users of the street, to the human perception of physical space, and in general to the quality of human-scaled urban design. With sufficient spacing between vehicles, or by keeping platooning to a maximum length, pedestrians will be ensured safe and frequent crossing opportunities as well as visual connectivity with their surroundings to safeguard human comfort. Correct spacing will prevent community fragmentation in districts and neighbourhoods, especially those with wide or busy streets, and will help to improve the sociability of places. Traffic signals will remain essential to allow pedestrians to cross higher speed and busier roads. Although CAVs could theoretically allow pedestrians to cross streets at will, if CAVs platooning is allowed this will either be impossible or difficult, and at least on busy streets crossing points should be provided. However, the design of such infrastructure will need to be radically reconceived to provide visual and sensory signals to pedestrians without the need to do so for vehicles. In the long term, road signs for cars could disappear and the number of street signs for pedestrians could be reduced, redesigned or supported by other technological advancements.

Design liveable streets

CAVs streets in urban cores should be designed primarily with the interests of pedestrians in mind, as social spaces where people can meet, and should have a shared space approach that will greatly reduce the demarcations between vehicle traffic and pedestrians. When vehicular movement consists exclusively of CAVs, pavements as they exist today may change as there will be no need to discourage drivers from parking or driving on pavements. Where possible streets should be redesigned at the same grade as pavements, without kerbs, or with a minimum kerb height (2.5 to 5 cm). However, kerbs may still be required to help visually impaired people navigate while new technologies are developed to support them more adequately. CAVs should be limited to a speed that does not disrupt other street users and demarcation of the space of different users could be achieved by rich contrasts in the physical environment including surface tactility, materials, patterns, colours, the enhancement of sound (especially once combustion engines are phased out), and other sensory clues such as bollards and landscape elements. The need for drop-off and pick-up space will increase exponentially and in urban cores this could mostly be provided along kerbs on new streets in a public and shared type environment, and could serve multiple addresses and different types of users throughout the day and night. These spaces will have pre-set or real-time maximum waiting time limits regulated (and potentially charged) by municipalities. For passengers, this could be up to 5 minutes during peak times.

Regulate access
Over time, only shared CAVs should have access to urban cores and should be allowed a maximum speed of up to 20 mph in local streets.[5] Outside core areas, during the transition phase, freeways or urban boulevards could have dedicated lanes for high-frequency CAVs separated from non-automated traffic.

The 18-metre rule compromise
In older cities, such as London, the limited width of the streets will require compromises to these principles. Many important city centre corridors are narrow,[6] as little as 18 metres (59 feet) building to building, meaning there can only be two lanes of traffic in each direction (or one lane in each direction and parking at the kerbside) plus a pavement on each side. With a view to maintaining bicycle connectivity alongside CAVs, bicycle lanes should be installed in both directions, alongside a lane in each direction for CAVs. However, two lanes of traffic in each direction will not be possible on narrow streets and a compromise will be needed. If the road is part of a strategic network with important implications for the overall capacity, then during the transition phase the layout should include mixed lanes with traditional cars and CAVs, thus compromising CAV benefits. Apart from these strategic roads, the street design should always favour people and bicycles, at the expense of vehicles.

KEEP IT LEGIBLE

The street has evolved over centuries and is the lens through which all city dwellers experience and interact with the world, ranging from the everyday experience of the residential street, to the heightened enjoyment of a bustling boulevard. Urban designers should be careful not to lose this understanding and experience of streets in the rush to embrace the new technology. Street design is nuanced according to local situations and preferences but the overarching language is the same the world over. Designers should take this into account, and help ensure that people will continue to be able to read and understand the street environment regardless of the technology within it. There could also be an opportunity to enhance aspects of street design, develop design approaches that better respond to pedestrian and cyclist needs and find ways to make CAVs behaviours understood by people.

Pedestrian traffic lights

Traffic lights will no longer be essential to manage traffic flow for CAVs, but on busy road corridors they will still be important for pedestrians and cyclists crossing the dominant flow. If CAVs drive bumper-to-bumper, traffic lights will still be needed to regulate flow and to provide gaps for people to cross easily. However, there may be scope to change their form and the user experience for pedestrians. For instance, signal poles and heads could become redundant and in some instances crossing points could be delineated on pavements and road surfaces using special tactile illuminated surfacing that is activated when pedestrians stand on or near it, with in-built auditory mechanisms that also communicate to hand-held or wearable devices.

Courteous crossing behaviour

Pedestrians rely heavily on eye contact with drivers and cyclists when crossing on crosswalks, as well as crossing streets away from formal crossing points. CAVs will be able to sense when pedestrians are in front of them and stop safely, however there will also be the need for CAVs to indicate their intention to stop when people are waiting to use a non-signalled crossing. This could be some form of visual or auditory signal (or both) on the car itself to show it acknowledges the presence of other users, to make them aware of its presence and to communicate the give way to them. Ensuring that CAV 'platoons' are limited to a maximum length and ensuring physical gaps in flow will also allow pedestrians to cross between them.

People-focused wayfinding

CAVs present an opportunity to remove much visual and physical clutter from street environments, including signage and road markings related to vehicular speed limits, stopping restrictions, traffic updates or directional information. However, elements that make streets legible for pedestrians and cyclists, including directional and other wayfinding information, can be retained and should be enhanced. As signs meant for drivers progressively disappear, and with them restrictions on street furniture positioning and requirements on clear visual lines, many opportunities will open up for new streetscape design and innovative walking and cycling wayfinding.

Natural legibility through design treatments

In addition to physical wayfinding elements it is important to maintain a language of street design in terms of materials that communicate visually and physically where pedestrians can expect CAVs to be. On roads with higher speeds and volumes of traffic, this is likely to mean continued use of traditional streetscape. On quieter local streets with a lower speed limit, a more pedestrian-focused approach should be used, with CAVs instructed to use defined vehicle paths through the street environment and pedestrians and cyclists being given priority through design treatment.

SHARE CAVS, SHARE STREETS

The use of CAVs on city streets should always be underpinned by the principle of sharing the street environment and infrastructure with other road users. The ethos of sharing should also be promoted in terms of CAV ownership and usage. In this way space can be maximised for people, enhancing liveability and activity on city streets.

Shared, not private CAVs

The space gains for pedestrians and cyclists may only be achievable if a shared-ownership model of CAVs is promoted or imposed. Private ownership will only increase the congestion and vehicle dominance we already see on city streets today.

All CAVs should be accessible

All shared CAVs within urban cores should be designed to be able to carry people with special requirements. Also, when servicing a person with limited mobility, they should be automatically allowed within certain areas that are normally precluded to general traffic. The technology should be about enabling people and providing support to those who need it the most.

Multi-functional kerbside areas

Space on city streets is currently highly regulated, with specific allocations for different types of parking, loading, and taxi bays. CAVs can facilitate the flexible use of such spaces and on-street parking should not be required if shared-use CAVs circulate and head to an off-street depot when not in use. Specific kerb space should be allocated for drop-off and pick-up to ensure that the frequency does not become a pedestrian barrier or undermine cycle safety. When space is needed for loading or servicing, a special restriction should be communicated to general CAVs, and the space cleared for registered delivery/servicing vehicles.

Multi-functional bus stops

Buses are expected to evolve as well, and should share infrastructure as much as possible with CAVs. Buses require long lengths of streets for stopping. Assuming that buses will be autonomous, it would mean that bus stop spaces could be used more efficiently. For instance, when no bus is due for 5 minutes, the space could be used for drop off or pick up by shared-ownership CAVs. In this way, bus stop waiting areas could become multi-functional and be used for shared CAVs services as well.

Consider options for time management on very constrained streets

In 'variable-geometry streets' traditional cars could gain access to the street or certain lanes only within specific time frames or traffic conditions and following a defined direction (one-way systems may also change over time).

REALLOCATE SPACE

Cities have the opportunity to reconsider how their street hierarchies function as part of movement networks. To maintain a pedestrian-first approach to street design CAVs should not be allowed to dominate city streets. Conversely, in order to ensure that CAVs will operate as efficiently as possible, CAVs and manual vehicles should not mix on all streets. Space allocation should be considered at a macro level, taking a layered network approach in order to allow all different modes to operate.

Integrated network planning

A multi-layered network approach is required, looking at city, district, and block scales to create connected routes for all modes. Taking a multi-layered approach entails identifying strategic routes for CAVs, manually-operated vehicles, buses and bicycles. These will not necessarily be on the same roads; routes for manually-operated vehicles, for example, should be limited to essential strategic routes. On some streets CAVs may be prioritised, but with separate cycle lanes to maintain bicycle connectivity, on others pedestrians and bicycles may be prioritised by allowing no other vehicles. However, pedestrian access should be maintained on all streets to an appropriate level given local demand and movement patterns.

Get the priorities right

Sustainability and liveability objectives still require a clear hierarchy between modes to promote the principle of *pedestrians first, then bicycles, then public transportation, then CAVs,* and *lastly manually-operated vehicles.* This hierarchy should be used for the design at the network level as well as at the street level, while always taking into account the local context.

Promote the macroblock (and its variations)

One of the best expressions of a layered network approach to city movement planning today is the macroblock. This approach works particularly well in grid-based systems such as the 'superblock' project in Barcelona.[7] A higher order of streets creates the macroblock, providing for movement of all types of road users around its perimeter. Within the macroblock, vehicle through-traffic is limited to spine streets. Other streets running off the spine streets are for pedestrians and bicycles first, or are 15 mph zones for access to local destinations.

PHASE OUT CARS

For CAVs to create a positive paradigm in cities, steps will be required to phase out manually-driven automobiles through regulation and design. However, there will necessarily be a transition period where both vehicle types will be on the road. During this period measures should be taken to prioritise CAVs.

Old cars on main roads only

To accommodate some level of accessibility the main roads in city cores should continue to host traditional cars, ideally in dedicated lanes. However, these should be for moving only, with stopping on such roads not allowed. Manually-operated vehicles should only be able to stop off-road in dedicated multi-storey parking garages that are redesigned as interchange hubs, providing access to CAVs, bicycles, and public transport. From these drivers will be able to cover the last mile of their journey on foot or by switching to another mode of transportation.

Over time manually-driven vehicles should be banned from city cores

CAVs should then be progressively allowed access only when shared and travelling at a maximum speed of 15–20 mph on local streets.

Deliveries and servicing

Streets will still need to accommodate essential service vehicles, such as for garbage collection or for delivery of goods to and from businesses and residents. Garbage collection vehicles will also be CAVs, accessing designated on-street stopping places or designated shared waste facilities located at specific points within the urban environment (for example one per city block). Delivery vehicles will also be CAVs, with coordinated intra-urban services operating to a delivery hub with larger vehicles. Smaller local vehicles would then transport goods to and from the delivery hub to locations in the area.[8]

Emergency service priority

Streets will still need to be designed to accommodate emergency vehicles, which will be operated manually for the foreseeable future.[9] By removing other manually-operated vehicles from all but main roads there could be a benefit for emergency vehicle access. By their connected nature, CAV movements will be able to be overridden by an emergency vehicle requiring access through a street, so that all other vehicles could stop or move aside in a much more coordinated manner. As they are manually operated, emergency service vehicles will still impose minimum spatial standards as today.

ENABLE NEW ARCHITECTURE

With the progressive introduction of CAVs for urban mobility, there will be a need for new kinds of building types to respond to emerging lifestyles and to the shifting paradigm between movement and place in cities, as with the introduction of the automobile at the beginning of the 20th century. CAVs will need to be stored, serviced and charged and while some of this can take place in decentralised facilities, there is likely to be insufficient space to address demand in dense urban areas, especially at peak times. If the aim is to reduce the amount of traffic and related carbon footprint created by CAVs, a new generation of mobility hubs (*NEMOHs*) will be required.

NEMOHs
would help to manage traffic pressure on the road network, would reuse land and create synergies with public transport rather than having the two modes conflicting with each other. CAVs could potentially undermine public transport as is currently the case with the private automobile, and in some cases with on-demand services like Uber and Lyft, with implications of further congestion and inequality. Therefore, NEMOHs will have to be combined wherever possible with existing and planned train, bus and subway stations. CAVs should access

the upper floors (or underground) while the ground floor of these hubs should function like a modern railway station, combining retail and leisure uses. These hubs should also include bicycle vaults, bicycle surgeries, bicycle and scooter sharing, and Click and Collect or small decentralised logistic facilities. CAV parking and waiting stations in urban cores will be then deployed as a network of hubs with different roles that have to be regulated and coordinated by local municipalities. In most cases, this infrastructure type will be built and operated by the private sector, or through Public Private Partnership (P3).

Other than accommodating CAVs and their needs such as waiting, parking, recharging and servicing, NEMOHs should first and foremost serve the users through their urban journey. They should provide people with different mobility options, services and shops, goods delivery and pick-up, and a comfortable place to wait, relax and socialise – a 'third place'. A typical hub could be a multi-level construction above or below ground. Most of the levels should be dedicated to CAV self parking. This will be with a parking efficiency higher than today's valet system, as CAVs will not require door-opening space, allowing them to occupy parking spaces that could be up 15 per cent smaller and could increase the capacity of a typical parking lot by more than 60 per cent.[10] Parking levels would typically not be accessible to the public, and some levels would be dedicated to goods or bicycle storage and servicing. Eventually, some of these hubs will also provide 'taxi drone' type services on their roofs and upper levels, for core-to-core and hub-to-hub passenger mobility and local delivery of packages. The ground level will be a new kind of public space – or 'privately-owned public space'[11] – or 'third place', with an active frontage, providing the main interface with the city and the users. Most of the drop-off and pick-up area will be at the ground level, part outdoor and part indoor. Large canopies will provide shelter for waiting users or those who have been dropped off.

Assuming that eventually urban cores will be accessible only to shared CAVs, parking needs will drop drastically. The provision of on-site parking spaces, as currently required for developments, should be ultimately revised in zoning codes, planning policies and building regulations. Existing and new constructions will no longer require provision for minimum parking on site. Instead, new constructions will have a maximum parking allowance or no parking allowance at all. Minimum kerbside requirements will still be needed for service and loading spaces and in some cases for drop-off and pick-up within the property. Most of the existing parking structures or parking spaces in buildings will be converted to other uses or could become NEMOHs, containing service centres and parking areas for CAVs waiting to be dispatched to users.

Over time different types of NEMOHs will be put in place as adoption accelerates, such as major, neighbourhood and remote type NEMOHs. These will eventually create a hierarchy, providing an organised structure to integrate CAVs and could phase out manually-driven vehicles from urban centres in the short term and be used to regulate traffic in the long term. The strategic conversion of current parking infrastructure to major hubs could create employment opportunities, cater for interchange needs and offer communities a hub for leisure and retail opportunities. Secondary NEMOHs will enable demand for CAVs to be satisfied by providing locations for vehicles to wait until needed.

As more and more vehicles on the road become hybrid and electric, the petrol station as we know it will decline. The demand for city and town centre surface and multi-storey car parking is also expected to diminish, as vehicles either park themselves in more remote locations or pick up the next passenger. Redundant or soon to be redundant automobile infrastructure will then provide the perfect opportunity to deploy the foundation of CAVs infrastructure. This could play a key role in the new mobility context, as it is already accessible and located at strategic points in cities and is owned by a mix of private and public interests. In coming years, public agencies will have the opportunity to deliver parts of the new CAVs infrastructure to test its functionality and effectiveness, and also its acceptance by the public. By using existing assets, there is a chance to lead an urban evolution and create a framework for private companies to work within. Reclaimed sites could be used to provide NEMOHs and also be developed to provide more housing, mixed-use developments, or parks to improve the social health and well-being of existing residents, while improving the ecological value of an area. Redevelopment and change of use of these sites will need to be considered very carefully, as a subsequent provision of CAVs infrastructure may be more difficult and expensive due to the scarcity of space in successful urban centres.

As linear kerb space becomes more valuable than parking lot space, building layouts should become more efficient with on-site parking progressively disappearing along with ramps and turning heads (a space where vehicles can manoeuvre in order other to turn). Layouts can also be adjusted to re-establish the primacy of the route from front door to the street, making it direct, legible and accessible. This would enable lobbies of larger buildings to evolve into more comfortable areas, facilitating social interactions among users waiting for CAVs. Existing buildings can be retrofitted, with car parking converted to alternative uses such as workplaces, leisure space or delivery hubs, with the potential to decongest streets. Service areas in buildings will also become more compact thanks to the manoeuvrability of CAVs and the automated coordination of timeslots. Buildings will also have the opportunity to be designed with a higher proportion of active frontage since the need to access service yards and on-site parking will drastically decrease.

CAVs offers the opportunity to radically review the way we organise urban logistics as well as how we service cities.[12] A network of neighbourhood logistic centres could be implemented to manage the last mile of deliveries using CAVs. These centres could work in synergy or be part of the larger mobility hubs system. Hubs could also host storage spaces for servicing equipment and provide decentralised production facilities such as 3D printing services to manufacture on-the-spot materials and tools, or replacements for machinery. Tradespeople could start using local hubs as a base for their operations, reaching their clients by CAVs or by bicycle and receiving the required equipment on demand via CAVs from the local centre. Click and Collect points could work as ATMs, being distributed and possibly integrated with mobility/delivery hubs. These local distribution centres could multiply in cores and should be publicly regulated and planned.

MAKE IT RESILIENT

The CAVs revolution also provides a unique opportunity to retrofit cities to become more environmentally sustainable and resilient. As evidenced by daily news, this issue is becoming increasingly urgent, with climate change and urbanisation exacerbating existing infrastructural systems, and proving them to be both too rigid and inadequate for the challenges ahead. As a consequence urban ecosystems are suffering.[13,14] Space gained by switching to a CAV-only fleet could be significant but there is a risk that without taking firm action through design, space gains could be lost to an increased volume of vehicles. If city cores are designed for shared CAVs only, allocation of space could be more radical, with major gains for pedestrians and public life through the building of denser, greener, and ultimately more sustainable cities.

Turn parking into parks

Giving priority to shared-use vehicles in city cores means that only shared CAVs would be allowed on local streets. Therefore, only limited drop-off and pick-up space will be required and a reduction in on-street parking could be achieved. Repurposed parking space provides an opportunity to reallocate street space, widen footways, install bicycle facilities, add greenery and 'blue' or water-based infrastructure, and create new pockets of public space. These changes should benefit both people and local environments.

Vast tracts of public and private space could be transferred from car use, with off-street surface parking converted to a system of parks connected through ecological corridors to existing green areas and to larger corridors to the countryside. Such an approach would enable urban cores to gain new

FIGURE 3.1
REDUNDANT CAR INFRASTRUCTURE.
A MULTI-STOREY CAR PARK IN
SHOREDITCH, LONDON

FIGURE 3.2
BOULEVARD IN THE TRAPEZE DEVELOPMENT,
PARIS, WITH AMPLE PAVEMENTS SERVING
BUSINESS ENTRANCES AND GENEROUS CYCLE
LANES AND GREENERY

ecological benefits and would facilitate natural flows and movements across the city. A system of parks could protect potential aquifers, support biodiversity, provide stepping stones for species and be used as a system of hydrological sponges against flooding.

CAVs should help to build healthy cities

The mobility benefits of CAVs are significant, but they could also attract more people to use them for first and last mile trips, replacing a walk or cycle ride to local public transport stations. We should be mindful of this and use the opportunity CAVs present to continue to build attractive, comfortable and safe local walking and cycling networks to support active travel behaviours. Additionally, CAVs will need to have appropriate speed limits in city cores for safety reasons but also to make walking and cycling an attractive alternative and to reinforce a high-quality environment for pedestrians.

WIN FAST (WHERE THE OPPORTUNITY ALLOWS)

The major challenge for cities in introducing CAVs will be to guide the positive transformation of environments by fostering the acceptance and public use of this new technology while ideally discouraging single-occupancy private automobile use. It will be important to get this transition right. The risk of creating another difficult legacy as seen in the case of the automobile is real, and quick and successful adoption will be important to secure the benefits of CAVs. Just as CAVs learn by driving, cities can learn to use driverless cars. This is why CAVs approaches should be tested as soon as possible by using driverless car regulated trials on streets – not only CAVs testbeds but also

real-life pilots focusing on how CAVs can improve cities and people's lives. Pilot areas might be corridors or entire city quarters, and it will be extremely important to implement pilot projects where these can maximise their chance of success and their impact in terms of public benefits. *A quick win*, whether in city centres or in the suburbs, will be to showcase the new way of life and boost public acceptance.

Start from the centre

The two case studies presented in this book propose a first implementation of CAVs starting from areas within the city cores of London and Los Angeles. Jumpstarting a broader CAVs strategy will involve demonstrating successful pilot projects in highly public areas where environmental gains are most needed.

Tighten the knots

CAVs may also provide an argument for densifying suburbs and helping to make public transport viable in areas where currently it is not. In this direction, it will be important to implement quick-win projects that promote CAVs as a microtransit option to complement walking and cycling for first and last mile access to mass transit. ▪

PART 2

||||||||||||||||||||||||||||||||||||||||||||||||||||||||||||||||||||||||||||||||||||||||||

VISIONS FOR
THE FUTURE

With cities, it is as with dreams: everything imaginable can be dreamed, but even the most unexpected dream is a rebus that conceals a desire or, its reverse, a fear. Cities, like dreams, are made of desires and fears, even if the thread of their discourse is secret, their rules are absurd, their perspectives deceitful, and everything conceals something else.

Italo Calvino, *Invisible Cities*, 1972

Chapter 4

LONDON VS LOS ANGELES: TWO CASE STUDIES FOR TWO VISIONS

How can we expect people to get out of their cars when their cars are much nicer spaces then sidewalks?

William H. Fain, *If Cars Could Talk: Essays on Urbanism*, 2012

The recommended urban design principles and underlying frameworks discussed have been tested on two areas in the central urban cores of London and Los Angeles, drawing parallels between the two. As already stated, urban cores represent the places where a first deployment of CAVs should occur because it is these areas where, if transition is managed well, the majority of urban design gains could be achieved. Secondly, these two specific cores present most of the typical characteristics that can be found in similar urban developments in Europe and America, providing a test ground for ideas that could be transferable to other cities and contexts.

The two global cities are emblematic within the paradigm of Western societies. The car-based pattern of development took place in both but while Los Angeles, a relatively new

American city, was free to adapt to the needs of the car, London's historic constraints led to a different incarnation of the model. In both cases, the effect was radical. Once again, urban environments in Los Angeles and London are at the forefront of a revolution that will unfold globally. But while metropolitan London today presents a multimodal approach to its mobility needs,[1] metropolitan Los Angeles relies on the automobile for about 90 per cent of commuter trips[2] and car culture is part of its DNA. Most other world cities fall between the two 'extremes' presented in London and LA. The two case studies suggest how the CAVs revolution could prompt a rethinking of these very different types of city core, as prototypes of what could be done in other American and European cities, and identify strategies that could be transferred to other places.

FIGURE 4.1
A DATA SNAPSHOT OF THE
TWO URBAN CONTEXTS

CITY DATA FROM LONDON AND LOS ANGELES		
	LONDON[3]	LOS ANGELES CITY[4]
POPULATION	8.8 MILLION (2016)	3.9 MILLION (2016)
SURFACE (HA)	157,200	121,211
DENSITY	55.9 INHABITANTS /HA	32 INHABITANTS /HA
CARS PER CAPITA	29%	51%
ROAD NETWORK	30,417 MILES	6,680 MILES
TRIP BY MODES OF TRANSPORT		
TRIPS BY PUBLIC TRANSPORT	45%	5%
TRIPS BY CAR, TRUCK OR VAN	31%	85%
TRIPS WALKING OR CYCLING	23%	4%
OTHER	1%	6%

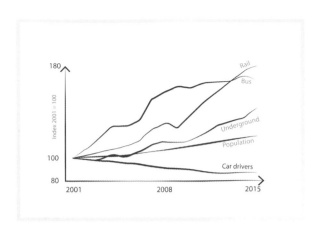

FIGURE 4.2
GROWTH IN JOURNEY
STAGES ON SELECTED
MODES, 2001–15, BASED ON
TRAVEL IN LONDON REPORT
9, MAYOR OF LONDON

LONDON IN CONTEXT

As London approaches the end of the second decade of the 21st century, its population is at an all-time high and is expected to reach the 10 million mark before 2030.[5] The economic success of the capital coupled with the tenacious protection of its Green Belt[6] has fostered a densification process, especially in inner London, which started in the 1990s.[7] The correlated growth of the commuter belt has absorbed the migration of Londoners taking refuge from unaffordable house prices, and now expands to the entire South East of England.[8]

These changes have placed increased pressure on London's infrastructure, and the current Mayor is responding to the challenge with a spatial strategy that prioritises growth around public transport hubs and a softening of density limits in exchange for higher percentages of affordable housing.

A number of high-profile infrastructure projects are currently being delivered, however the majority relate to rail infrastructure upgrades, extensions or new lines.[9] This approach reinforces

London's renowned tradition of public transport excellence and, coupled with the established trend of disaffection with the private car, is confirming Londoners' proclivity for multimodal trips.

In terms of road infrastructure the picture is rather different, with Transport for London and highways departments of local municipalities[10] trying hard to make the most of the existing network, which is not expected to be upgraded significantly, especially in terms of capacity. If anything, capacity is contracting[11] as the city policies are giving more weight to environmental considerations and reallocating road space to sustainable modes of transport and green infrastructure.

Congestion on London's streets is getting worse,[12] despite a clear cycling renaissance,[13] the decline of driving licences among millennials[14] and the steady decrease in car ownership.[15] Ultimately a large component of congestion relates to the rigidities and physical limits of the road network created by London's urban structure, which over centuries has resisted 'modernisation' attempts by eminent architects including Sir Christopher

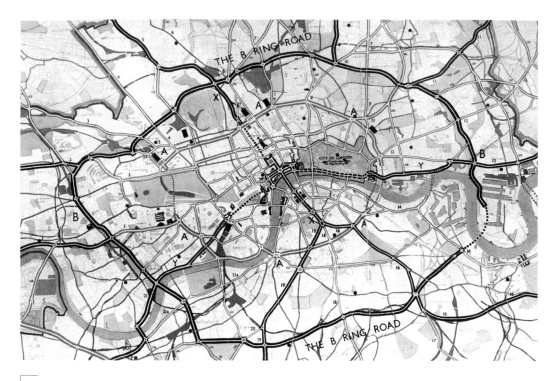

FIGURE 4.3
ROAD PLAN FROM THE COUNTY OF LONDON
PLAN, 1943.

Wren and John Nash. In central London, the combination of the nature of urban fabric and the concentration of jobs creates a fierce competition for space. The constraints of the network are exemplified by the inner ring road, which includes the busiest roads of the capital yet is made up of long stretches of narrow sections: at certain points Pentonville Road, for example, is 20 metres (65.6 feet) wide, just enough for three lanes of traffic, a narrow cycle lane plus the pavements. Similarly, Commercial Street is less than 18 metres (59 feet) wide at certain points, and Tower Bridge Road is 18 metres (59 feet) wide.

Many of London's streets struggled to accommodate non-motorised traffic during the Victorian age, so it comes as little surprise that attempts to comprehensively redesign the road network intensified in the 'century of the car'. One of the most notable endeavours was by Sir Leslie Patrick Abercrombie. His idea was to reshape the spatial structure of London on a series of spokes and rings of roads, however only fragments of this grand plan were built after World War II.[16]

Today, London remains a city with a strategic street network formed by a series of historic and mixed-used high streets. This feature of the city is in some senses a curse as it is fundamentally unfit to deal efficiently with high volumes of traffic, but is also an asset in terms of character and, most importantly, it performs as a social and economic hub.

The spatial limitations of London's streets and of the form of its road network make the task of improving environmental performance and air quality and expanding the provision of amenity space, while increasing or at least preserving capacity, very daunting.

In this context, the opportunities created by shared CAVs should be looked at very closely, as they could be a strategic tool for solving some of the current issues of London's streets. Provided that the deployment of CAVs will be able to deliver the expected benefits, they could make a significant difference to London if rolled out on a large scale. In this sense accelerating adoption in London is vital.

This new addition to the city also presents a fundamental risk because hurried implementation may erode, at least in the short term, the quality of London's built environment, in a way not dissimilar to what happened in the 1960s. London is still decluttering its streets from redundant or over-engineered car paraphernalia, once enthusiastically deployed to facilitate the presence of the car in the city while 'protecting' the other street users. But the last thing city dwellers need is to witness yet another wave of road segregation, especially barriers between pedestrians, cyclists, traditional cars and CAVs, all intended to enable the latter to use the streets more effectively.

LOS ANGELES IN CONTEXT

With the revolution of the private automobile after World Wars I and II, Los Angeles witnessed a rapid and unprecedented transformation that is still underway. At the turn of the 20th century, the city was a relatively small and peripheral centre within the US and counted a population of approximately 100,000 people within the original 28 square miles.[17, 18] The first gasoline-powered vehicle appeared on the streets of Los Angeles in 1897[19] and by 1932 the city had grown to 450 square miles, increasing to almost 2 million in population by 1950.

Today, the City of Los Angeles has a population of over 4 million over an area of 472 square miles while Los Angeles County counts over 10 million people in an area of 4,084 square miles.[20, 21] The Los Angeles-Long Beach-Anaheim Metropolitan Area is the second largest metropolitan area in the US with a combined population of 13.3 million people in an area of 34,135 square miles.[22] By 2025, the Greater LA area is expected to be the densest urban area in the nation with an estimated 6,450 people per square mile and a projected total population of nearly 15.7 million.[23] According to the Southern California Association of Governments (SCAG), 90 per cent of daily commuter trips in the region are currently made by private automobile.[24] Lewis Mumford called Los Angeles the 'reductio ad absurdum' of the cult of the car – a city hijacked by the false promises of the motor age.[25] Apart from Detroit, there is no American city more identified with the automobile than Los Angeles. Automobiles are part of the city's milieu. Movies like *Crash* (2005) and the award-winning *La La Land* (2016), or the work of contemporary local artists such as Carlos Almaraz and David Hockney, all testify how car-culture has become part of the city's genetic order or DNA; an auto-oriented landscape for an auto-oriented population.[26]

In the last 20 years, the region has made relatively modest progress in addressing Modernist mistakes of car-led city planning, dismantling some barriers created in the name of traffic efficiency and desegregating some uses. Even today, reconfiguring streets and urban freeways, or at least some of them, is considered unfeasible or is approached with extreme caution in LA, while in US cities such as San Francisco, New York City, Boston and even Dallas projects of this type have been carried out successfully for years.

Policies from organisations such as Complete Streets and Vision Zero[27] have received fiery push-backs in this culture of car-led planning. The recently adopted Measure M, a ballot measure to fund transportation projects in LA County, is still a pro-car measure for the most part, even if it will foster the development of alternatives to the automobile.[28] One of the most innovative documents recently released by the city, *Urban Mobility in the Digital Age*, provides a roadmap for the city's transportation future. The report addresses the city's plan to combine self-driving vehicles with on-demand sharing services to create a suite of smarter, more efficient transit options.[29] However, substantial work remains to make this vision a reality and to establish a reliable multimodal transportation system for the region while avoiding complete gridlock.

This condition of almost total dependency on the automobile for urban mobility is fairly typical within the context of US cities, with Los Angeles, Detroit and Houston being among the primary examples of car-centric urbanism worldwide. One of the much-presaged benefits of the CAVs revolution will be more efficient traffic operations leading to the increased carrying capacity of existing road infrastructure and fewer parking requirements, with the potential to reallocate road and parking space away from vehicles to other uses.[30] With the ever-increasing levels of traffic congestion that the Los Angeles urban region is experiencing and which, according to SCAG, with population growth will increasingly experience,[31] there will be a temptation to simply use the space 'gained' to accommodate more vehicles. City makers should instead take this opportunity to instigate a paradigm shift and break the cycle of automobile dominance on Los Angeles streets. The CAVs revolution is a great opportunity to use this 'extra space' to retrofit the urban region in a more climate and context conscious and sustainable way. With the advent of CAVs, planners and stakeholders involved in city making will have a once-in-a-lifetime opportunity to reconsider how Los Angeles' streets function as part of a movement network. To introduce a hierarchy to a historically undifferentiated (to other than cars) gridiron plan, a multi-layered network approach at different scales will be required to create connected routes for all modes. CAVs will most likely constitute a mix of shared service and private ownership, and private CAVs should progressively be banned from accessing urban cores within the region as private ownership will only promulgate the congestion and vehicle dominance that we see in Los Angeles today. In order to promote multimodal and pedestrian-first environments, only shared CAV uses should be allowed access to city centres. Could the downtowns of Los Angeles, Santa Monica, Pasadena, Long Beach or Burbank be reorganised in such way?

SITES SELECTION CRITERIA

The areas selected to test the CAVs strategy in these two vast cities are the Shoreditch neighbourhood in London and the Financial District in Downtown Los Angeles. These particular areas were selected for two reasons. Firstly because urban cores today represent the places where a first deployment of CAVs will occur, and where the majority of urban design gains could be achieved. Secondly, these areas present most of the typical characteristics that can be found in similar urban developments in Europe and America, thus providing a test ground for ideas that could be transferable to other cities and contexts.

Even though they present very different urban morphologies, as evident in the figure-ground diagrams below, these two areas also have a series of relevant comparable characteristics:

- **Urban cores:** The study areas are part of the cities' urban cores and are not located in suburban locations.

- **A quarter of a mile or 5 minutes' walk (0.5 mile square area):** The characteristics of the areas are meaningfully represented within a 5 minute walk from a central point. This is a manageable size for the study in practical terms as well as being large enough to observe different transportation modes and patterns.

- **Streets and blocks:** The two areas represent typical street and block conditions.

- **Mixed-use:** The two areas contain mixed land use, including commercial use, and a variety of building typologies capturing the typical complexity of the urban settings.

- **Presence of public transport:** The areas are both served by public transport and showcase a comprehensive set of modal conditions.

- **Close to the citywide urban network:** The areas are close to a freeway of a major artery of access to the city. ■■■

FIGURE 4.4
LONDON VERSUS LOS
ANGELES – A LAYOUT
PLAN OF THE STUDY
AREAS

Chapter 5

LONDON

This should be the promise of urban life:
the city's diversity of urban life becoming
a source of mutual strength rather than
a source of mutual estrangement and
civic bitterness.

Richard Sennett, *"Civility" Urban Age*, 2005

PRACTICAL STUDY AREA: SHOREDITCH

Shoreditch is emblematic of London's complexity in many ways. It features a distinctive range of uses from employment to residential and from leisure to culture. Its urban fabric features an eclectic mix of building types from different eras. Shoreditch is also subject to an intense process of densification typical of many inner cities across the globe, with multiple developments completed in recent years and many still in the pipeline. As parks and public squares are scarce, its public space provision relies heavily on the street network, which is under pressure to respond to amenity, environmental and business demands in addition to performing its movement

functions. The eclectic nature of Shoreditch is manifest in every aspect of its character, including its irregular layout, which reflects historic development. The area emerged organically as a ribbon development along the old roman road (now the A10) that connects the City (London's central square mile) to East Anglia. Shoreditch hosts a mix of commercial, retail and residential uses as well as leisure activities. The street grid remained irregular until the Victorian age, when incremental improvements (first to Commercial Street, then to Great Eastern Street) and associated slum clearance improved the situation somewhat by creating an orbital route from City Road to Whitechapel, and then further south to the London Docks and Tower Bridge, bypassing the City.[1]

FIGURE 5.1
SHOREDITCH STUDY AREA LOCATION PLAN

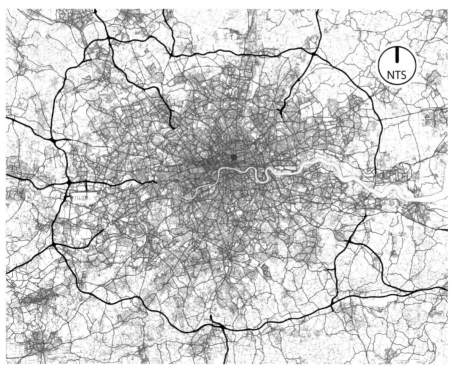

Except for the redevelopment of the sites bombed during World War II and the much more recent wave of new tall buildings creeping northward from the City, the urban fabric of Shoreditch has seen little change since the Victorian age. Inevitably, the street network has struggled to cope in terms of capacity and environmental quality in the age of the automobile. In the 1960s the main streets of the study area (Shoreditch High Street, Old Street and Great Eastern Street) were turned into an intimidating mile-long gyratory system with sections of up to four lanes of one-way traffic.

Despite these changes, the impact of vehicular traffic remained significant in this area, partly due to the introduction of the Congestion Charge in 2003.[2] The boundary of the toll area in this part of London coincides with the Inner Ring Road including Great Eastern Street and Old Street. As a result, the main arteries are under further pressure as vehicles try to avoid the charge zone during the day. In this sense its location adjacent to the City, in addition to being an economic blessing for the regeneration of Shoreditch, also creates a number of infrastructural challenges. Away from the main arteries, the network is made up of small streets and tight

FIGURE 5.2
STREET PLAN OF SHOREDITCH, LONDON

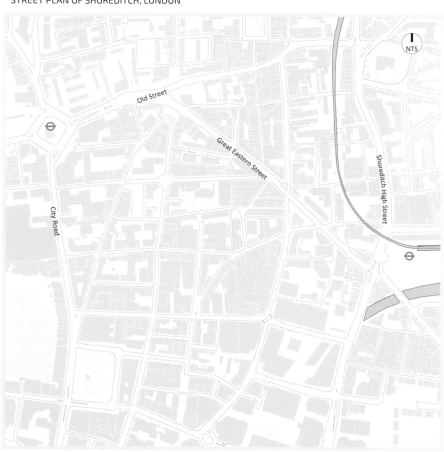

FIGURE 5.3
TYPICAL STREET CONDITION IN SHOREDITCH

intersections and is characterised by a complex one-way system that does not, however, discourage drivers and has also devoted a substantial amount of public space to street parking. The complexities of the Shoreditch context are challenging but nevertheless not dissimilar to those of many other neighbourhoods in London. In a way Shoreditch reflects the conditions of many other European city districts. There is a final and important reason for looking at Shoreditch and it is a symbolic one: it is the home of London's start-ups and creative and digital technology scene. What better testing ground for the possible introduction of CAV technology? If the CAVs strategy works in Shoreditch then it could be used as a blueprint for the whole of London and beyond.

SPACE BATTLES

If – or when – CAVs are accepted by the public, they will not replace the old automobile overnight. It could take up to a generation[3] to decommission the whole fleet of manually-operated cars and replace it with driverless ones. The 'short-term' scenario of needing to accommodate both types of vehicle on London's streets is in fact likely to be prolonged. However, the speed and scale of adoption will be critical factors if CAVs are to deliver their promises, so the local authorities will be under pressure to redistribute the little road space available. In general, CAVs will benefit from keeping the mix of old and new vehicles to a minimum in the same space. If this does not happen, then CAVs could turn into a glorified electric car e-hailing service[4] and this would eradicate many of the advantages they offer. Demographic pressure and current infrastructural rigidities will demand bold choices, especially as CAVs will claim street space at the expense of traditional cars without bringing tangible benefits in the short term. How do you switch to CAVs in a constrained place like Shoreditch if banning traditional cars altogether in central London is not yet a feasible option, especially politically?

The response is to implement dedicated CAV Corridors (C-Corridors). These would be able to complement the traditional vehicular network without overlapping with it. In the beginning they are envisaged as a separate layer used by traditional cars and public transport and meeting the rest of the network only at key points. These joints would host a new generation of mobility hubs created to facilitate the switch to or from CAVs and long-distance public transport. They would intercept drivers who could park their cars and either use a driverless car or walk or cycle to complete their journey. NEMOHs[5] could also be the places where CAVs would be stored and serviced and would be located along the edges of central London in dedicated new structures or retrofitted parking garages.

Given the reduced amount of space available and the risk of bringing traffic to a complete halt with the addition of CAVs to general traffic, initially there would only be a small network of C-Corridors enabling fast, convenient movement from central London to the edges of the inner city and to orbital routes allowing east-west and north-south movement. The streets between these corridors could be progressively closed off to general traffic except driverless cars. Starting from the centre of the capital, this network could grow to cover the whole city over two decades or so as public acceptance becomes the norm and Londoners increasingly enjoy the benefits.

FIGURE 5.4
GREAT EASTERN STREET
IN SHOREDITCH. THE
THOROUGHFARE IS PART
OF THE LONDON INNER
RING ROAD DESPITE ITS
CONSTRAINED WIDTH

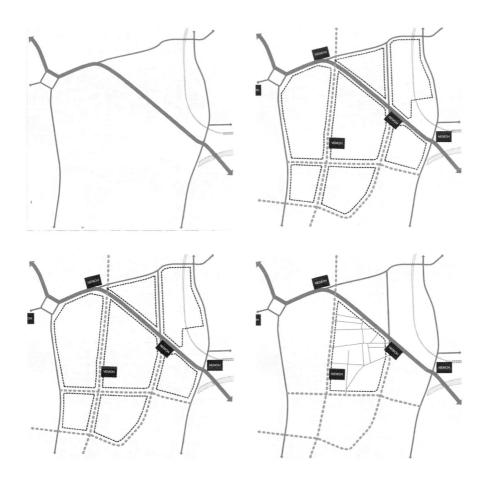

CELLS AND CORRIDORS

CAV-only lanes on some sections of the strategic network would be possible, however the benefits to CAVs are likely to be minimal as bottlenecks in the network would cause vehicles to mix frequently. Where CAVs share lanes with buses, the space and capacity gain would also be limited especially if the buses are not driverless and connected; there could be no reduction in the width of the lane nor in the distance between CAVs and manually-driven buses. And there would be the risk of merely compromising the performance

of public transport, effectively lowering its rank in the movement hierarchy. The reality is that the strategic network, including the section running through the heart of Shoreditch, will probably remain unaltered for at least another decade except in the unlikely circumstance of a drastic reduction in its capacity. The only noticeable change would be the appearance of CAVs between buses, cars, white vans and cyclists and... more traffic.

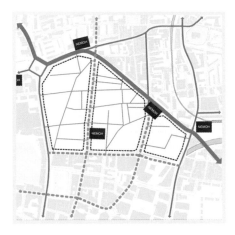

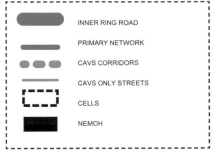

INNER RING ROAD

PRIMARY NETWORK

CAVS CORRIDORS

CAVS ONLY STREETS

CELLS

NEMOH

FIGURE 5.5
PROPOSED PHASED
INTRODUCTION
OF CAVS IN SHOREDITCH

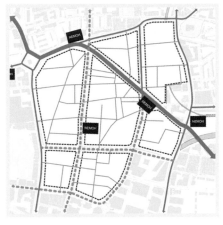

The biggest opportunities in the short term lie off the main roads in corridors dedicated to CAVs and cyclists, especially in inner London, and in promoting the shift to shared CAVs at dedicated stations distributed around the ring road. These corridors could be designed and managed with the objective of bypassing the surrounding congestion and eventually promoting the spread of a wider network free from traditional vehicles. The neighbourhood 'cells' defined by the corridors would be progressively closed to general vehicular traffic except for CAVs completing routes with origins or destination addresses within the cells. Local streets then turn into shared surfaces or streets with a narrow carriageway, from 2 to 4 metres (6.5 to 13 feet). These streets would still have to cater for emergency vehicles but these can be accommodated beyond the constrained limits of the formal carriageway, provided the adjacent areas are free from obstacles. The rest of the space would be used for cycle infrastructure, additional pedestrian space, blue and green infrastructure, spill-out space for businesses and HODOs and service bays for driverless vehicles.

C-Corridors and areas within the 'CAV cells' will have a low maximum speed, for instance of 15 mph, for two reasons:

- Safety in case of accidents (shared CAVs will use lighter vehicles).

- Getting the priorities right: the more sustainable options are still walking and cycling. CAVs must be more convenient than cars but less convenient than other sustainable means of transport. For instance, 15 mph is only marginally more the average speed of cyclists in cities,[6] and would enable positioning of cycling as a quicker and healthier way to move around the city.

On local streets, we anticipate a further process of decluttering as vehicular paraphernalia becomes obsolete. Space dedicated to CAVs will be differentiated through the use of materials, colours, or light but full separation from the other modes of transport on the road should be avoided. These local streets will see the biggest gains in terms of public realm and sustainability targets. It is the public right of way, not the road network, that will benefit through the retrofitting of former parking spaces for alternative purposes.

As Londoners become more familiar with CAVs and become used to interacting with them, the amount of demarcation and traditional street furniture can progressively be toned down or altered into different forms. It would still be very useful, for instance, for pedestrians and cyclists to know where to expect to see a CAV, or where space needs to be kept free for deliveries. Parking would still occur, but less frequently and in a more intelligent way with either pre-reserved space or in a flexible manner by, for example, morphing to loading bays when required. Some level of physical demarcation ought to stay, if only as a back-up system should the technology fail.

Some locations along kerbs will support charging facilities but these will eventually be replaced by induction technology,[7] which could reduce clutter and potentially even maintenance costs. Ideally, charging would occur over stretches of roads while vehicles are in motion, rather than kerbside where vehicles must be parked over extended periods of time. In any case, the majority of charging should occur in the mobility hubs and in CAVs servicing facilities.

The progressive demise of manually-operated cars will be facilitated by creating a series of mobility hubs along the strategic network. This would be achieved by retrofitting existing filling stations and multi-storey parking garages and by building new ones. Secondary hubs created in central areas would be able to deal with peak demand. Within a few years, people driving from afar would be able to drive their cars then drop them off at a mobility hub and shift to public transport, shared CAV, or bicycle.

THE FUTURE STREETS OF SHOREDITCH

How could Shoreditch evolve following the application of the CAVs strategy for London?[8] The answer lies in an approach to reconfiguring streets into types that respond to local idiosyncrasies.

There is no standard Shoreditch street, instead there is a full range of typologies from alleys to residential streets, from major arteries to mixed-use local distributors, from shared surfaces to roundabouts. Despite this diversity there are also common denominators, especially the limited widths and poor provision of green infrastructure. In addition, the eclectic mix of buildings defining Shoreditch's character are an important element of how the

streets work, given the thin transitional spaces between their entrances and the pavement, which result in frequent encroachments on to the public space outside shops, bars and other buildings.

MAIN ROADS

Key routes such as Old Street, Great Eastern Street, and Shoreditch High Street, criss-cross Shoreditch and form part of the London Inner Ring Road. The widths of these streets are as narrow as 16 metres (Old Street in Hoxton), 17.5 metres (Great Eastern Street) and 18 metres (Shoreditch High Street) and allow traffic in two directions, high-frequency bus services, a number of pedestrian crossings and traffic lights at short intervals. In addition to the strategic road network, Shoreditch

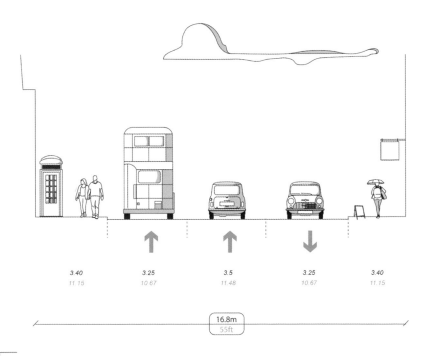

| 3.40 | 3.25 | 3.5 | 3.25 | 3.40 |
| 11.15 | 10.67 | 11.48 | 10.67 | 11.15 |

16.8m
55ft

FIGURE 5.6
CITY ROAD, EXISTING CONDITIONS OF A LOCAL CONNECTOR

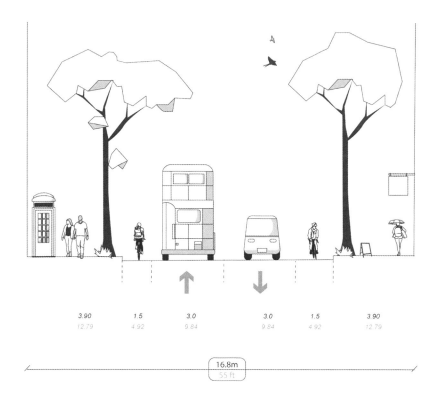

| 3.90 | 1.5 | 3.0 | 3.0 | 1.5 | 3.90 |
| 12.79 | 4.92 | 9.84 | 9.84 | 4.92 | 12.79 |

16.8m
55 ft

FIGURE 5.7
CITY ROAD, PROPOSED MEDIUM-TERM CONDITIONS OF A LOCAL COLLECTOR

features local connectors such as the southern section of City Road, Hackney Road and Bethnal Green Road. The width of these roads is also extremely constrained as, in addition to hosting bus routes without continuous bus lanes, they also support two-way traffic and parallel street parking.

In the short term, these important streets will broadly remain the same but incremental change may include the possibility for shared CAVs to use bus lanes at certain times of the day, the creation of a limited number of flexible spaces serving as HODOs and reserved loading bays for driverless vehicles. However, allowing CAVs and buses to share the same lanes on the strategic

road network would compromise the public transport system, which should continue to feature higher than CAVs in the movement hierarchy as it is inherently more efficient. In the medium term, while the strategic road network may remain unchanged, local collectors could become bus- and CAV-only roads. These streets would feature a carriageway of 5.5 or 6 metres (18 to 20 feet),[9] plus 3 metres (10 feet) for cycle lanes, mixing the two types of traffic (especially if buses also become driverless) and reallocating space to pavements and shared drop-off areas. At selected locations, the gain of space will allow the introduction of additional trees as well as loading bays and HODOs.

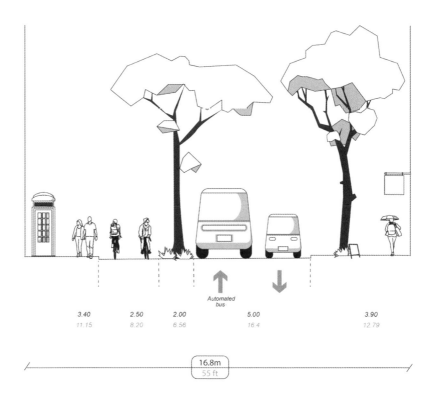

3.40   2.50   2.00   5.00   3.90
11.15   8.20   6.56   16.4   12.79

Automated
bus

16.8m
55 ft

FIGURE 5.8
CITY ROAD, PROPOSED LONG-TERM CONDITIONS OF A LOCAL COLLECTOR

C-CORRIDORS

C-Corridors would link together a series of local roads to create continuous routes off the main vehicular network. The current layout of Shoreditch's streets typically reserves a large proportion of the road space for parking. C-Corridors consist of a narrow roadway – between 3.5 and 4.5 metres (11.5 to 15 feet) depending on local conditions and overall street width and directionality – flanked by a two-way cycle lane of at least 2.5 metres (8.2 feet). No parking would generally be allowed but HODOs could be arranged on the opposite side of the cycle lane, given sufficient street width. Where

extra space can be gained, pavement widening, street planting, rain gardens and other types of sustainable urban drainage systems (SUDs) could be used to enhance the characteristics of the C-Corridors.

Where the C-Corridors meet the rest of the network, which is accessible to manually-operated cars, priority would be given to cyclists and CAVs via stop lines for the side streets. C-Corridors would be clearly defined, especially in the early stages, with specific colours, materials and streetscape treatments for the purpose of both visibility and promotion (see Figures 5.9–5.11).

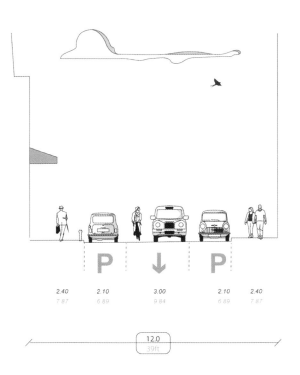

FIGURE 5.9
EXISTING LOCAL STREET

| 2.40 | 2.10 | 3.00 | 2.10 | 2.40 |
| 7.87 | 6.89 | 9.84 | 6.89 | 7.87 |

12.0
39ft

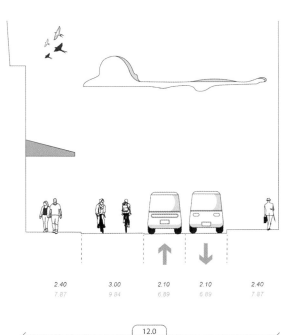

FIGURE 5.10
LOCAL STREET RECONFIGURED
AS A C-CORRIDOR

| 2.40 | 3.00 | 2.10 | 2.10 | 2.40 |
| 7.87 | 9.84 | 6.89 | 6.89 | 7.87 |

12.0
39 ft

FIGURE 5.11 VISUALISATION OF C-CORRIDOR (BEFORE AND AFTER)

LOCAL STREETS

Different approaches have to apply to local streets, depending on their context (street width, bordering land uses, frequency of access points, servicing requirements, and so on). However, in the long run local streets would ideally feature three common elements:

- A complete ban of traditional cars. Private vehicles, whether cars or CAVs, would have to be stored at dedicated facilities outside the CAV-only zones or further out at the edges of the city.

- A maximum speed limit of 15 mph.

- Conditional access – in other words only CAVs originating or completing a trip at an address within the local street network will be allowed access. However, in exceptional circumstances the network could be opened to CAVs provided there is an organisation capable of monitoring and coordinating the traffic flows for the whole network.

In addition local streets would have to be designed to maximise opportunities for sustainable urban drainage systems and planting, and to integrate frequent HODOs (at least one per block to maximise accessibility). Segregated cycle lanes would not be required given the low speeds and the limited amount of traffic.

The design variations should include at least three main configurations:

Shared surfaces
As with traditional shared surfaces, there would be no differentiation between pavements and roadways. This type of arrangement would be used in quieter streets, especially residential ones, where CAVs traffic is unlikely to be in conflict with pedestrians and cyclists or where disruptions to CAVs movement would be deemed acceptable. While a section of 3.5 metres (11.5 feet) can be used for the circulation of bicycles, CAVs and emergency vehicles, the rest of the street space will be retrofitted as pedestrian space and, importantly, with SUDs and other green infrastructure. These green stretches would also be punctuated by HODOs and loading and service bays for protracted parking requirements. The number of bays would depend on the land-use mix and local density. To minimise their visual impact and to maximise environmental gains of green stretches of ecological corridor, bays will in general not be clustered. No other parking would be allowed.

A 3.5m carriageway for CAVs as well as for cyclists
This would be an appropriate treatment for the narrowest street, with CAVs allowed in one direction only. Provided that CAVs are communicating across different service-provider platforms, the one-way street would also change the direction of the flow (flexible one-ways) to respond to CAVs routing as vehicles will know if others are already using the street and in which direction. Directionality could also respond to the level of traffic congestion during the day. As CAVs are assumed to have a width below 2 metres (6.5 feet), the extra 1.5 metre (5 feet) is provided for cyclists. In general, this total width would also help to address the requirement of current emergency vehicles to be able to pass through.[10]

The rest of the street should be reserved for pedestrians, greening/SUDs, HODOs and loading bays. As the greatest opportunities to reallocate space lie in local streets, it is interesting to assess the indicative amenities

FIGURE 5.12
LOCAL STREET CURRENT
CONDITIONS

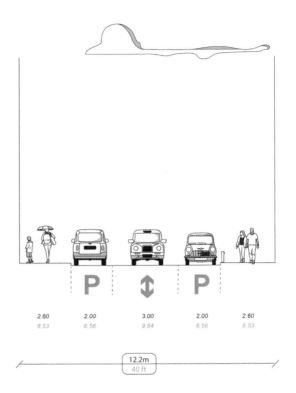

P ↕ P

| 2.60 | 2.00 | 3.00 | 2.00 | 2.60 |
| 8.53 | 6.56 | 9.84 | 6.56 | 8.53 |

12.2m
40 ft

FIGURE 5.13
LOCAL STREET PROPOSED
CONDITIONS

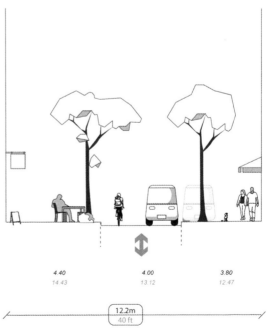

| 4.40 | 4.00 | 3.80 |
| 14.43 | 13.12 | 12.47 |

12.2m
40 ft

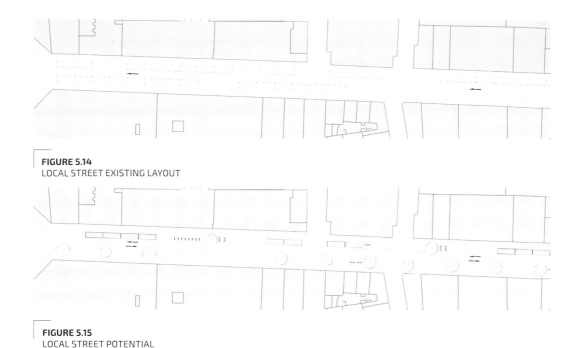

FIGURE 5.14
LOCAL STREET EXISTING LAYOUT

FIGURE 5.15
LOCAL STREET POTENTIAL
LAYOUT WITH CAVS

gained. Amenity gain means the space that can be devoted to non-movement functions, such as SUDs, parklets, tree planting, business spill-out spaces and public seating areas.

Testing the approach

The intermediate approach described above was tested by applying it to a 200 metre (656 foot) section of Leonard Street in Shoreditch. Plots fronting the street are typically mixed-use comprising retail, office and residential uses and have an indicative plot area ratio of 2:5. These are common conditions in London, with many high streets featuring similar characteristics. According to the simulation it would be possible to reclaim up to 25 per cent of the street space (building line to building line) for public amenities (see Figure 5.15). This is equivalent to

an area of around 600 square metres (6,500 square feet)[11] featuring 40 trees. If applied to all local streets in the Shoreditch study area, this could mean creating an urban forest of approximately 1,200–1,300 trees able to absorb at least 7 tonnes of CO_2 per year[12] and an overall provision of green space or other amenity space equivalent to four football fields.

Despite the number of assumptions affecting these calculations,[13] the results are very positive. While CAVs won't transform urban streets into bucolic country lanes, an initial assessment suggests they provide the opportunity for a dramatic improvement of London's credentials in terms of sustainability, microclimate, air quality, amenity and ultimately liveability.

STREET TYPE RELATIONSHIPS

In the creation of a multi-layered network, it is important to address the issues relating to the interface between different types of roads. This is particularly relevant for pedestrians and cyclists, to help them understand the environment they are navigating, but also for car drivers to clearly understand the areas they are accessing and if they are allowed to do so. Both C-Corridors and car-free neighbourhoods (cells) could feature threshold treatments – but not in the dramatic colour schemes such as those delivered by the London Cycle Superhighways. Instead it will be important to showcase and promote the corridors while being sympathetic to context. Enforcement against manually-operated cars entering restricted zones could take place via electronic gates and non-obtrusive road signs to warn drivers. The entry points could feature roadway narrowing and, where possible, small public spaces. As C-Corridors become priority routes in the short and medium term, there would also be opportunities to stress their status in terms of special streetscape details. Finally, these streets could feature continuous pedestrian pathways along their whole lengths and possibly bespoke lighting.

NEW TECHNOLOGY, NEW BUILDINGS

Shoreditch's streetscape would not be the only visible change prompted by CAVs, as the new vehicles would be accompanied by NEMOHs. These would take some of the spatial arrangements of familiar transport infrastructure types, like train stations and parking garages, but would ultimately result in different types and variations depending on their location and function. In fact, NEMOHs in Shoreditch will be of two main types:

1. Major hubs

The hubs along the strategic road network would inevitably be busier and larger as they facilitate the change from private vehicles or public transport to CAVs. Where possible these would be integrated with existing underground and train stations to maximise their potential and promote the use of public transport. They would have a prominent street frontage and would target users with integrated retail and leisure opportunities. Click and Collect services would be available and would eventually offer the possibility to pre-load CAVs with goods ordered by the single user. For this reason, hubs would include small logistics centres to unpack deliveries for the last mile. Obvious locations for Shoreditch would be Old Street Roundabout and especially the Bishopsgate Goodsyard by Shoreditch High Street Station.

Surviving filling stations in the city such as those on City Road and Old Street could be redeveloped to accommodate these hubs. Over time, as manually-operated cars decrease in number, more and more space in the hubs would be reallocated in favour of CAVs.

2. Neighbourhood hubs

A second type of NEMOHs would be located near the proposed C-Corridors to balance the load on the CAV network and to respond to peaks in demand, particularly near the city where the fluctuation of demand will be more pronounced. The creation of some of these facilities would be opportunistic, for instance achieved by retrofitting existing multi-storey parking garages, while others will be new planned facilities on available plots, or more likely mixed-use plot redevelopments integrating residential, office and retail to improve viability. These hubs would be smaller and have less presence on the street, with minimal street frontage as most of the CAVs hosted will leave the structure to reach clients elsewhere.

FURTHER IMPLICATIONS FOR THE BUILT FORM

CAVs will trigger the emergence of new types of facility but also promote changes in established typologies. Space-intensive functions such as service yards could become more efficient depending on the level of automation in terms of loading and unloading and would require less space as CAVs will be able to manoeuvre more effectively. The need for plots for car parking will reduce drastically. Parking policies will also evolve. Parking minimums will be replaced by maximums[14] and will directly relate to the amount of kerb space allowed for drop-off or HODOs size within walking distance. In this scenario on-plot drop-off areas would be the exception rather than the rule and HODOs would be charged per use by local authorities to recover the municipal revenue losses from parking fees and tickets.

As on-site parking garages and service yards decline, buildings would increasingly have less 'back-of-house' space and more usable frontage. The relationship between public space and the private sections of buildings would improve while the importance of lobbies and transition areas would increase, as is already happening for other reasons in Shoreditch. Local studios and shops have already started a process of blending functions and spaces, promoting the creation of diffuse semi-public space at ground floor, with public pavements and small yards and with coffee shops and bars becoming retail spaces and meeting rooms.

As CAVs make more street space available in quieter, less polluted and more attractive local streets, and as they improve the relationship between streets and buildings with the reduction of service areas or blank frontage, businesses would have further incentives to spill out into the public realm to boost street life. Finally, the spread of CAVs would stimulate Shoreditch's further densification, as the last remaining surface parking garages and gas stations will become available for redevelopment.

Greening and public realm improvements will be a common thread of the CAVs strategy and will respond and adapt to the individual character of each street and building with a wide palette of interventions. Environmental improvements will create a green network that will radically change the ecological performance of the neighbourhood in terms of drainage, air quality, amenity and biodiversity.

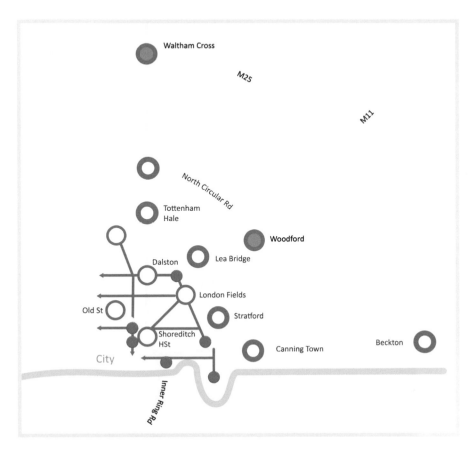

In the next 20 years Shoreditch will still be characterised by the buzz and traffic of its three busiest thoroughfares that define the 'Shoreditch Triangle', but outside these main roads the area has the potential to become a green neighbourhood taken over by pedestrians, cyclists and, along well-defined corridors, by CAVs. The substantial increase in amenities could make this part of London even more desirable for both residential and commercial uses and, unless the same logic is quickly applied to other areas of London, it could accelerate even further the gentrification processes at the eastern edge of the City.

NEMOH TYPES
- MAJOR URBAN EDGE
- MAJOR STORAGE/ SERVICING FACILITY
- INTERCHANGE HUBS
- LOCAL NEMOHS/ STORAGE FACILITY
- C -CORRIDOR

FIGURE 5.16
AN INDICATIVE NEMOHS SYSTEM. HUBS WILL HAVE DIFFERENT CHARACTERISTICS DEPENDING ON THEIR LOCATION AND ROLE. REMOTE HUBS WOULD STILL BE REQUIRED ON THE OUTSKIRTS OF LONDON WHEN C-CORRIDORS AND CAR-FREE NEIGHBOURHOODS SPREAD TO THE EDGES OF THE CITY. REMOTE HUBS NEAR MAIN ARTERIES (E.G. POTTERS BAR AND WALTHAM CROSS) WOULD CONNECT LONDON TO THE REST OF THE COUNTRY, AND COULD MORPH INTO INTERCHANGE STATIONS TO INTERCEPT LONG-DISTANCE DRIVERS SWITCHING TO CAVS TO COMPLETE THEIR JOURNEY TO SHOREDITCH OR CENTRAL LONDON

WHAT IF CARS WERE BANNED FROM LONDON'S STREETS?

To attempt to quantify the possible space gains resulting from the introduction of CAVs, the following simulation imagines their introduction in some London streets if traditional automobiles were banned. The three streets 'tested' are an existing mixed-use busy street in a central zone, an existing residential street with average access to the public transport network, and a residential street that is part of a new development in a suburban context with relatively poor access to public transport.

The simulation has been based on the calculations for two-way trips generated by the current land uses at peak hours. This is derived from the TRICs database, a tool regularly used by transport planners in the UK. A notional 20 per cent increase in demand has been assumed in order to anticipate a potential increase of trips reflecting the increased accessibility of vehicles (such as use by the disabled, teenagers and the elderly), and sets the hourly capacity of each CAV HODO bay at 15 vehicles per hour, with CAVs occupying space for hop-on/drop-off for a maximum four minutes. (This time allocation is the typical grace period before ride-hailing services start charging for waiting.) The rest of the time it is assumed that CAVs will be in circulation or stored at NEMOHs.

THREE CAVS SCENARIOS

FIGURE 5.17
WHAT IF CARS ARE REPLACED BY CAVS

SCENARIO 1 – A ZONE 1 MIXED-USE LONDON STREET WITH VERY HIGH ACCESSIBILITY (PTAL 6B)

EXISTING SITUATION

100 RESIDENTIAL FLATS (80 ONE-BED AND 20 TWO-BED) ABOVE RETAIL (6,500M2) AND OFFICE SPACE (13,500M2), PLUS A 125-ROOM HOTEL

CURRENT PARKING PROVISION: 34 SPACES

POSSIBLE SCENARIO WITH CAVS

SAME LAND USE MIX AND QUANTUM.

PARKING BAYS REQUIRED: 5 PLUS 9 SERVICE BAYS

TOTAL: 14 SPACES

POTENTIAL SPACE GAIN: UP TO 59%

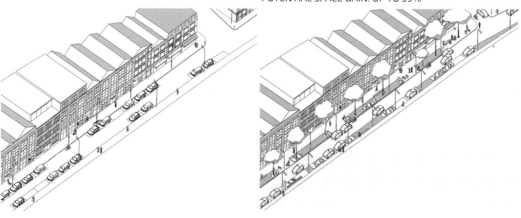

SCENARIO 2: A ZONE 2 RESIDENTIAL LONDON STREET WITH AVERAGE LOW ACCESSIBILITY (PTAL 3)

EXISTING SITUATION

98 UNITS (48 ONE-BED, 30 TWO-BED, 20 THREE-BED)

CURRENT PARKING PROVISION: 46 SPACES

POSSIBLE SCENARIO WITH CAVS

SAME LAND USE MIX AND DEVELOPMENT QUANTUM

PARKING BAYS REQUIRED: 1 PLUS 2 SERVICE BAYS

TOTAL: 3 SPACES

POTENTIAL SPACE GAIN: UP TO 93%

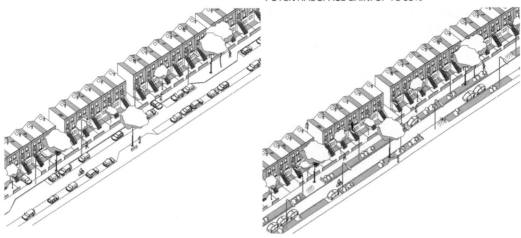

SCENARIO 3: A ZONE 4 LONDON RESIDENTIAL STREET WITH LOW ACCESSIBILITY (PTAL 1B)

EXISTING SITUATION

120 UNITS (30 ONE-BED, 60 TWO-BED, 24 THREE-BED AND 6 FOUR-BED)

CURRENT PARKING PROVISION 120

POSSIBLE SCENARIO WITH CAVS

SAME LAND USE MIX AND DEVELOPMENT QUANTUM

PARKING BAYS REQUIRED: 1 PLUS 2 SERVICE BAYS

TOTAL: 3 SPACES

POTENTIAL SPACE GAIN: UP TO 98%

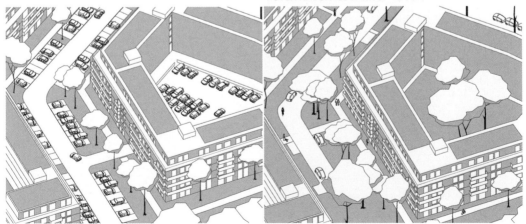

Space has also been allocated for servicing and delivery bays, but this is calculated according to current demand-generation models and assuming that the servicing is not CAVs. This is to reintroduce some spare capacity into the model, as the bays will be used quite inefficiently. It has been also assumed that these streets would be accessible only to CAVs.

For simplicity, carriageway dimensions have been maintained, as well as parking spaces of 2.5m by 5m. In the short term a substantial reduction of carriageway width is unlikely if the road has to continue catering for existing emergency service vehicles, refuse trucks and so on. Car parking dimensions are also likely to remain the same in the short term, given that CAVs technology at the moment is fitted into the current car chassis.

Finally, an important design element, relating to the even distribution of bays along streets in relation to block size and the location of entrances, has been disregarded. Considerations in the area should warrant the reintroduction of some redundancy, or extra bays, to improve the convenience and accessibility of CAVs.

There are, in short, a high number of assumptions. Changes to any of these variables would alter the outcome. Most importantly, nobody knows at this stage how CAVs will change attitudes towards travel in the city. For this same reason, these types of exercises are useful as we need to build working hypotheses and then test these in real-life urban environments. ■■■■

Chapter 6

LOS ANGELES

Come to Los Angeles! The sun shines bright, the beaches are wide and inviting, and the orange groves stretch as far as the eye can see. There are jobs a-plenty and land is cheap. Every working man can have his own house (and his own car), and inside every house, a happy, all-American family... Life is good in Los Angeles... It's paradise on Earth.

"Sid Hudgens" in the film *LA Confidential*, 1997

PRACTICAL STUDY AREA: DOWNTOWN LOS ANGELES

To showcase the possibilities of the CAVs approach in the City of Los Angeles, a 0.5 square mile area of the Financial District in Downtown (DTLA) was selected. As clearly evidenced in the famous and still relevant 1970 report *Concept Los Angeles: The Concept for the Los Angeles General Plan*,[1] the Los Angeles urban region can be understood as a stellar system of multiple centres, or cores, connected in different capacities through major corridors, with DTLA being the main core within the region. The ideas presented here for DTLA could be applicable to other LA cores as well.

Downtown Los Angeles's Financial District (see Figure 6.3) presents the typical undifferentiated grid pattern of streets and blocks that are common to almost every city core in the US. The grid plan – since its first use as the physical foundation for Philadelphia by William Penn in 1682[2] – has been used extensively as a structural framework in a number of American cities in every one of the 50 states.

The typical DTLA block size measures about 350 by 550 feet (107 by 168 metres) and the typical street section varies from 80–90 feet (24–27 metres). By contrast, the smaller New York City blocks in Midtown Manhattan measure 400 by 200 feet (122 by 61 metres) and have a typical street section of 60 feet (18 metres). These are potentially good dimensions for a walkable environment. Even with the substantial improvements of the last years, thanks to massive private real estate investments and the injection of more than 60,000 new

FIGURE 6.1
PROPOSED REGIONAL
CENTRES AND TRANSIT
BLUEPRINT FROM THE
CONCEPT LOS ANGELES PLAN

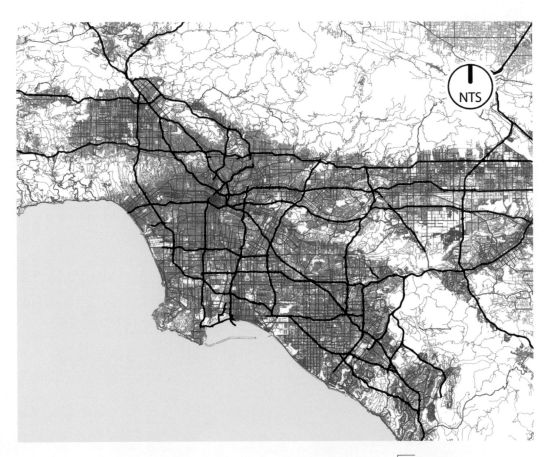

FIGURE 6.2
STUDY AREA LOCATION
PLAN – DOWNTOWN
LOS ANGELES

FIGURE 6.3
TRAFFIC DOMINATED ROAD
IN DOWNTOWN LOS ANGELES
WITH THE CHARACTERISTIC
ONE WAY SYSTEM.

residents,[3] years of car-driven policies and car-driven mindsets, shaped and still shape the unpleasant urban experience of DTLA. A one-way street system ensures access to freeways for fast-moving vehicular traffic while over-dimensioned parking standards and exaggerated perceptions of market demands also produce an over abundance of on-street and off-street parking spaces.[4]

According to a report published in the *Journal of the American Planning Association* in 2010, 14 per cent of the 200 square miles of LA County's incorporated land is dedicated to parking infrastructure, leading to a total of 18.6 million parking spaces. These break down into 3.6 million on-street and 15 million off-street – or about 3.3 parking spaces for every car, with

about a third of the off-street spaces being residential. DTLA is home to the greatest parking density, with some census tracts in the central business district with upwards of 260,000 off-street parking spaces per square mile, mostly packed into multi-level garages.[5]

It can be assumed that at least 70 per cent of a typical DTLA street is dedicated to the needs of cars and that, based on a study by Manville and Shoup,[6] more than 50 per cent of the DTLA area is dedicated to streets and to on- and off-street parking. Of the approximately 3,000 acres of land in DTLA more than 1,000 acres is dedicated to car uses. These numbers give a sense of the potential for a CAVs strategy to unlock the city in terms of recapturing this space, or at least part of it, for better uses.

FIGURE 6.4
DOWNTOWN LOS ANGELES FINANCIAL DISTRICT TYPICAL STREET GRID

FIGURE 6.5
DOWNTOWN LOS ANGELES TYPICAL STREET CONDITIONS

TYPICAL URBAN BLOCK DIMENSIONS 350 X 550'

OPEN SPACE

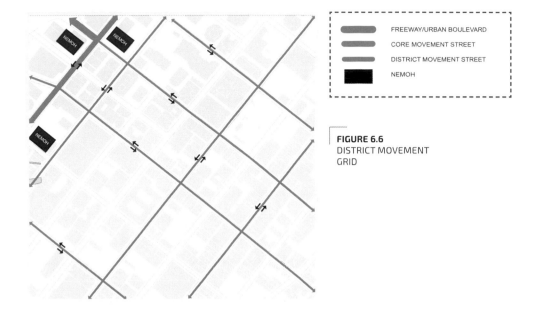

FREEWAY/URBAN BOULEVARD

CORE MOVEMENT STREET

DISTRICT MOVEMENT STREET

NEMOH

FIGURE 6.6
DISTRICT MOVEMENT
GRID

CAVS SUPERGRID

A CAVs strategy gives the opportunity to radically rethink the street hierarchy of urban cores, like DTLA, in terms of movement and place. Thanks to the potential abundance of new available space and to the redundant properties of the grid system (qualities that are not present in contexts like the one discussed in London), it is possible to imagine how in the near future urban cores like DTLA could be reshaped.

The following diagrams show the proposed CAVs strategy for DTLA, which consists of a long-term new four-level hierarchy for vehicular movement imposed over today's undifferentiated street grid. The hierarchy is made up of: *multi-way urban boulevard; core movement; district movement; and local movement streets*.

The efficiency of CAVs and increases in real estate value present an opportunity for urban freeways cutting through urban cores – like the 110 Freeway

in the diagrams – to be progressively decked over or reconfigured into multi-way urban boulevards with dedicated public transport space. This opportunity could help to reconnect and revitalise adjacent urban fabrics to regain a more human scale within the core, while carrying more people faster using the same or less space. Core movement streets would define the different districts within the core; district movement streets would define superblocks within districts; and local movement streets would provide fine-grain access and services to single properties.

When, or if, CAVs make up the majority of vehicles on the streets and reach Level 5 or full automation[7] (estimated by the 2040s in Los Angeles),[8] it would be advisable that the urban cores are made progressively accessible only to shared public or private CAVs systems, while access for manually-operated automobiles would be banned. Adopting a real-time congestion charge could regulate the amount of privately-

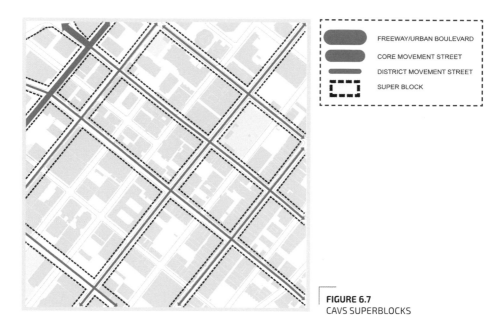

FREEWAY/URBAN BOULEVARD

CORE MOVEMENT STREET

DISTRICT MOVEMENT STREET

SUPER BLOCK

FIGURE 6.7
CAVS SUPERBLOCKS

owned CAVs reaching cores using the system of freeways and multi-way urban boulevards. If used, privately-owned CAVs would then have the option to drop off or pick up owners in dedicated mobility hubs, or NEMOHs, at the periphery of the core along the multi-way urban boulevards and then self-park in remote, less valuable locations. From mobility hubs, passengers would be required to shift to a shared CAVs system, use public transport, cycle or walk, to reach their final destination in the core. Despite the political difficulties, private CAVs movement would eventually be permitted only on core movement streets to access NEMOHs. Eventually, these hubs could also accommodate 'taxi drone' types of service for core-to-core and hub-to-hub passenger mobility.

The maximum speed allowed within the core for all vehicles would be 25 mph. District movement streets' maximum speed would be 15–25 mph, and for local movement streets would be

15 mph. Even if CAVs drastically reduce the risk of crashes and injuries by up to 90 per cent,[9] these limits would keep speeds down to a safe level for all road users while improving the quality of life on the street. It has been proven that lower speeds in fact reduce community fragmentation in districts and neighbourhoods,[10] and ultimately improve the *sociability* of places. District movement streets will accommodate CAVs, public transport and shared CAVs system services to connect different districts. All streets will be reconverted to a two-way system.

District movement streets define the superblocks. The dimension and configuration of each superblock varies according to local constraints and land uses, but the blocks generally define a specific sub-district and pedestrian-first zone, and foster particular neighbourhood identities.

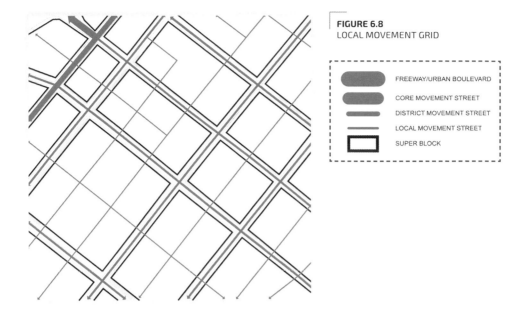

FIGURE 6.8
LOCAL MOVEMENT GRID

FREEWAY/URBAN BOULEVARD

CORE MOVEMENT STREET

DISTRICT MOVEMENT STREET

LOCAL MOVEMENT STREET

SUPER BLOCK

Local movement streets would have a maximum speed limit of 15 mph. Shared CAVs would be permitted access only if the origin or destination of the trip's address is on the street.

The reorganisation of DTLA's grid into district movement streets and superblocks will optimise vehicular movement and, by redesigning street sections, will allow the introduction of *an active grid layer* on local movement streets. The active grid layer connects superblocks in central areas with existing and new provision of open and green space. The active grid is thought of as a movement and place system for walking, biking and rolling, but also as a green/blue infrastructure for environmental sustainability and as a linear system of open spaces and urban parks for public enjoyment and sociability. The active grid will be a capillary system serving all areas within the core.

Urban streets are public spaces, and street intersections are primary public spaces. The efficiency of the new CAVs supergrid system, by reducing the space needed for vehicular movement and parking, would permit the redesign of intersections as meaningful public spaces for people, would add to the city's amenity offer and would improve the overall quality of the urban experience. Local street intersections in the middle of superblocks could be redesigned to become plazas, squares, or pocket parks, and will be the centre of social interaction for superblocks. DTLA currently has about 500 city blocks. Assuming a future configuration of approximately 125 superblocks and 125 local intersections, each occupying an area of 10 to 15,000 square feet, for a total of about 36 acres (with half of this space reclaimed as public space) DTLA could incrementally add about 18 acres of public and green space simply by reconfiguring this type of intersection.

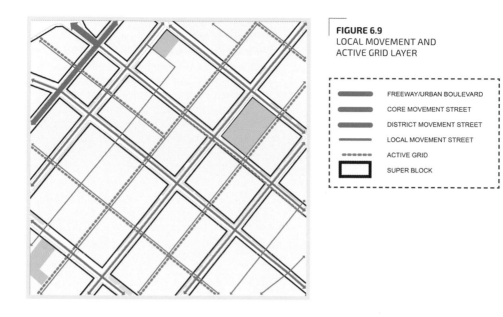

FIGURE 6.9
LOCAL MOVEMENT AND
ACTIVE GRID LAYER

FREEWAY/URBAN BOULEVARD
CORE MOVEMENT STREET
DISTRICT MOVEMENT STREET
LOCAL MOVEMENT STREET
ACTIVE GRID
SUPER BLOCK

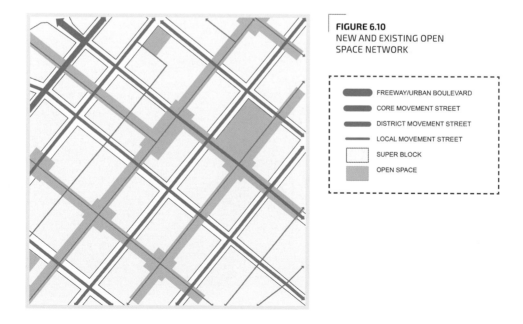

FIGURE 6.10
NEW AND EXISTING OPEN
SPACE NETWORK

FREEWAY/URBAN BOULEVARD
CORE MOVEMENT STREET
DISTRICT MOVEMENT STREET
LOCAL MOVEMENT STREET
SUPER BLOCK
OPEN SPACE

A RESILIENT URBAN FORM

The CAVs supergrid strategy would also be an opportunity to retrofit urban cores to be more sustainable and resilient. Assuming that in urban cores CAVs will be almost exclusively of shared use and multiple occupancy type (as latest market predictions indicate[11]) it could be assumed that by 2040 more than 50 per cent of the current parking space in DTLA could be redundant. Considering only surface parking – on-street and parking lots – this corresponds to a total surface area of 800 acres or more, the size of Central Park in New York.

FIGURE 6.11
A CONNECTED SYSTEM OF
MEDIUM AND LARGE PARKS

Taking advantage of this vast public and private space regained from use by cars, the active grid could be paired with a new system of parks, or green patches, connected through ecological corridors to the existing green areas and to the larger natural corridors outside the city. This would allow urban cores like DTLA to gain new ecological benefits by facilitating natural flows and movements across the urban area[12] and would improve climatic resiliency. It would also be an opportunity to provide a mix of urban parks, green areas and vegetable gardens as a form of civic amenity that is almost completely absent in DTLA today, and which represent an important element of the social infrastructure critical to the competitiveness agendas of cities worldwide.[13]

The system of parks also allows the protection of potential aquifers to support biodiversity, providing stepping-stones for species and offering a network of hydrological sponges against flooding. Ideally, these parks will have a minimum dimension of 9 acres – about two DTLA blocks or more – and will be placed at no more than 0.5 miles or a 10-minute walk from one another.

FIGURE 6.12
THE URBAN GRID VERSUS
THE ECOLOGICAL GRID

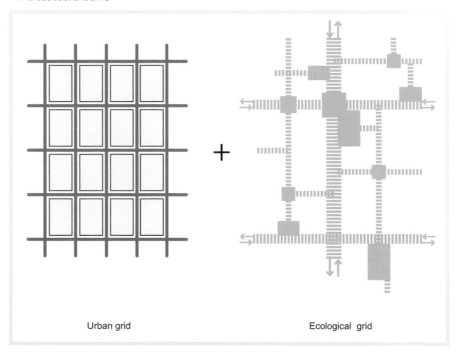

Urban grid   Ecological grid

This configuration maximises the combined cooling capacity of the system across the core and would help reduce the heat island effect of the surrounding environment – an issue that will only intensify in DTLA in the coming years. To understand the cooling potential of urban parks in contrast to the heat island effect, it is helpful to look at the research done on urban climate in relation to Tiergarten Park in Berlin by Stulpnagel, Horbert and Sukopp. Their study concludes that the larger the park, the greater the temperature difference compared to the built-up surroundings.[14] A passive energy strategy of this type could bring a noticeable temperature reduction to the surrounding areas. A 10-minute walking distance between parks also ensures that everyone living or working in DTLA has access to meaningful greenery within around a 5-minute walk.

Space in the parks could also be designated to contribute to the protection of water quality and to mediating urban watershed, with specially designed storm water ponds, for example, being used to control large volumes of water. In the design of this system of parks, each park site should be understood in terms of its relation to its watershed and other possible water sources. Ideally, to accommodate ecological concerns and to provide a valuable programme for the city, parks should be split into two zones: one being more 'naturalised' where most of the ecological issues can be addressed, and one being more heavily programmed to address the required uses of the local population.
Local streets would work as linear water

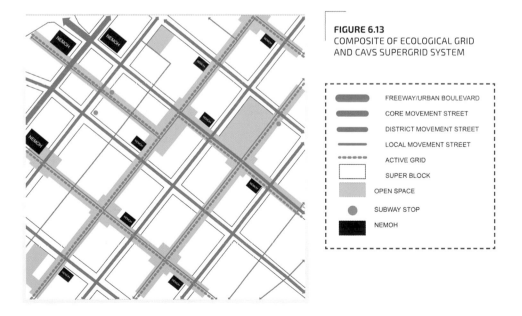

FIGURE 6.13
COMPOSITE OF ECOLOGICAL GRID
AND CAVS SUPERGRID SYSTEM

FREEWAY/URBAN BOULEVARD
CORE MOVEMENT STREET
DISTRICT MOVEMENT STREET
LOCAL MOVEMENT STREET
ACTIVE GRID
SUPER BLOCK
OPEN SPACE
SUBWAY STOP
NEMOH

retention systems and also as green and ecological corridors, permitting the dispersal of plants and animals. These corridors could play a key role in restoring and maintaining biodiversity and the continuity of ecological processes in a heavily modified environment such as DTLA.

Environmental quality is also directly related to tree cover in cities. Trees and woodland can provide great benefits to the urban environment by providing shade and beauty, and play a substantial role in the overall functioning of such a system of parks and corridors through providing ecological benefits and a cooling effect. The American Forest Association recommends that tree coverage in cities is up to 40–60 per cent under ideal conditions in forested states, is 20 per cent in grassland cities, and 15 per cent in desert cities like Los Angeles. These are baseline targets, with higher percentages possible through greater

investment and prioritisation.[15] The CAVs supergrid strategy would provide a great opportunity for a city lacking in woodland and green spaces to start implementing these policies, addressing habitat fragmentation in its core and, by doing so, improving its ecosystem values.

MOVEMENT AND PLACE

With new ideas and potential reconfigurations of the city's parts within an incremental strategy, DTLA streets could be redesigned following the CAVs supergrid strategy.

The diagram in Figure 6.14 shows a typical superblock configuration, consisting of four city blocks for a total dimension of approximately 680 by 1,300 feet (207 by 396 metres). It is served on the four sides by district movement streets and internally by two local streets. The longest dimension is about a 5-minute walk end to end.

FIGURE 6.14
TYPICAL SUPERBLOCK CONFIGURATION

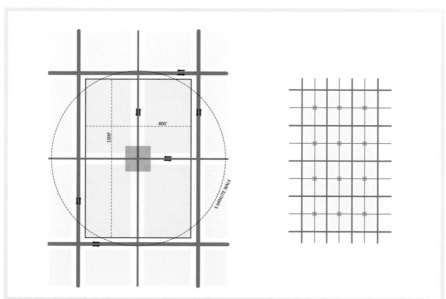

District movement streets

District movement streets will accommodate CAVS primarily in the form of public transport and shared services with the function of connecting the different districts within the core. Streets will be reconverted into a two-way system, and private CAVs movement will not be permitted on this type of street. Maximum speed limit for CAVs will be in the region of 15–25 mph.

The diagram in Figure 6.16 shows a current typical street section within our study area, in this case Olive Street. The building-to-building distance is about 85–90 feet (26–27 metres). Of this length, about 10–12 feet (3 to 3.6 metres) per side is dedicated to pavements – 24 feet (7.3 metres) total – and the rest of the 65 feet (20 metres) to one-way vehicular traffic, with street parking on both sides.

By using the CAVs supergrid strategy, Olive Street is reconfigured to become a district movement street. The diagram in Figure 6.17 shows how the street section is redesigned this way, essentially as a mini multi-way boulevard. The central portion of the street will accommodate shared CAVs through-traffic and public transport services. Because of CAVs efficiency, only one lane per direction is provided. Central lanes allow a maximum speed of up to 25 mph. Median strips on the two sides will provide space for pedestrians to comfortably cross the street at the ends and middle of the block and to sensibly reduce crossing distances. Medians are also used to allocate stop and waiting areas for public transport services, for green and blue infrastructures, and for

planting trees. On the internal sides, one lane for CAV movement each way will provide local access and service to single parcels, to drop-off and pick-up kerbside areas (HODOs), and to access the district's mobility hubs (NEMOHs) where CAVs could temporarily park, receive service, or recharge. The maximum speed on these lanes will be 15 mph and could be shared by cyclists. Pavements will be extended to a minimum of 15 feet (4.5 metres) where possible. This provides space for pedestrians to walk as well as sit and socialise, to improve the landscape and trees offer, and to add to the urban realm amenities.

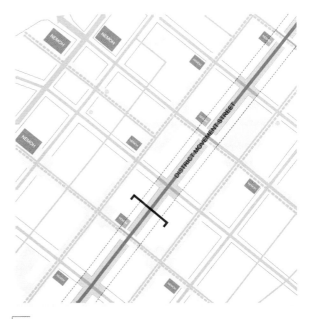

FIGURE 6.15
TYPICAL DISTRICT
MOVEMENT STREET

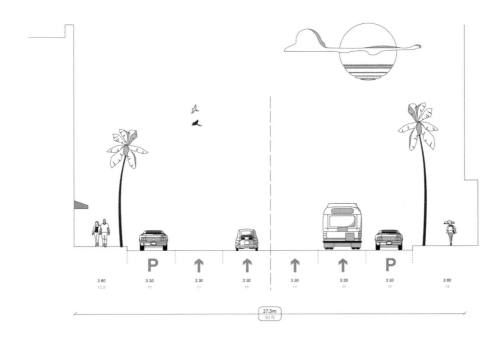

| 3.60 | 3.30 | 3.30 | 3.30 | 3.30 | 3.30 | 3.30 | 3.60 |
| 12.0 | 11 | 11 | 11 | 11 | 11 | 11 | 12 |

27.5m
90 ft

FIGURE 6.16 EXISTING CONDITION OF A DISTRICT MOVEMENT STREET

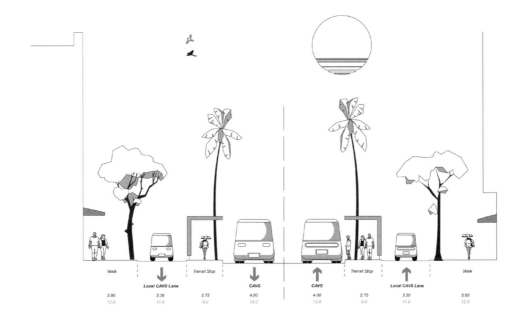

Walk	Local CAVS Lane	Transit Stop	CAVS	CAVS	Transit Stop	Local CAVS Lane	Walk
3.60	3.30	2.75	4.00	4.00	2.75	3.30	3.60
12.0	11.0	9.0	13.0	13.0	9.0	11.0	12.0

FIGURE 6.17 FUTURE CONDITIONS OF A DISTRICT MOVEMENT STREET WITH CAVS TRANSIT STOP

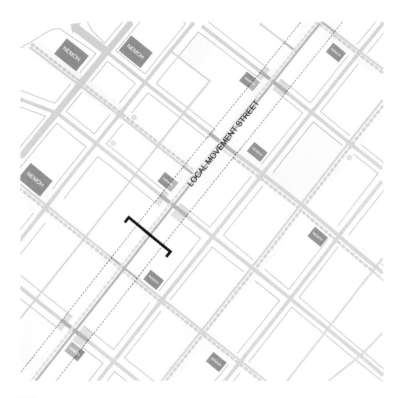

FIGURE 6.18
A TYPICAL LOCAL MOVEMENT STREET

Local movement streets

Local movement streets serve to guarantee access to single parcels within the superblock. These are advised to have a maximum speed limit of 15 mph and shared CAVs services that use them only if the origin or destination of the trip is on that street.

For a typical street section along South Grand Avenue (Figure 6.18), the building-to-building distance is about 85–90 feet (26–27.5 metres). Of this length, about 10–15 feet (3–5 metres) per side is today dedicated to pavements and the rest of the 65 feet (20 metres) to one-way vehicular traffic with street parking on both sides, in a similar arrangement to Olive Street. Recently, the City of Los Angeles has implemented a Complete Street scheme on this section of the street.

By using the CAVs supergrid strategy, this portion of Grand Avenue is reconfigured as a local movement street. With a maximum speed limit of 15 mph and with shared CAVs services allowed access only if the origin or destination of the trip is in that street. The street would be used to access single parcels, to reach a HODO, to use the new amenities of the enhanced public realm or simply to stroll along.

The central portion of the street accommodates shared CAVs local traffic. Because of CAVs efficiency, only one lane per direction would be necessary, and about 25 feet (7.6 metres) or less would be needed to accomplish this. Medians on the two sides, by reducing crossing distances, provide space for pedestrians to comfortably cross the street at multiple points along the block length. Medians also demarcate user areas, provide space for drop-off and pick-up, or HODOs, and for green and blue infrastructures and tree planting.

The internal sides of the street would be used to accommodate the active grid layer and the ecological corridor functions. The diagram in Figure 6.19 shows two cycle tracks of two lanes each, fast and slow, on the two sides that could be used for primary (commuter) trips and secondary (local) trips.

Pavements would be extended to a minimum of 15–20 feet (4.5–6 metres). This provides space for pedestrians to walk as well as sit and socialise in correlation with ground floor uses, and to add to the landscape and trees offer. Pavements and medians accommodate the linear storm-water retention system. The combined landscape and pervious surface treatment act as an ecological corridor to encourage the dispersal of plants and animals, while contributing to the reduction of the heat island effect.

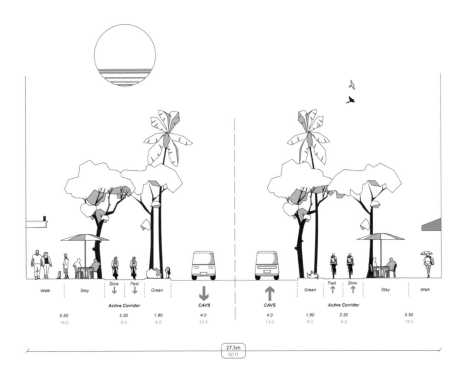

FIGURE 6.19 FUTURE CONDITIONS OF A LOCAL MOVEMENT STREET

Local intersections

Local street intersections in the middle of superblocks will be redesigned to become plazas, squares, or pocket parks and to provide new public amenities and cafes. These would be conceived as the centre of social interaction for the superblock.

The first diagram in Figure 6.20 shows how the intersection at Grand Avenue and 8th Street functions today. Both streets are one way, with three lanes each and on-street parking on both sides. A class II bike lane was recently added on Grand Avenue. Of the current 10,000 square feet (930 square metres) of the intersection area, about

FIGURE 6.20
LOCAL MOVEMENT STREETS INTERSECTION – PROPOSED

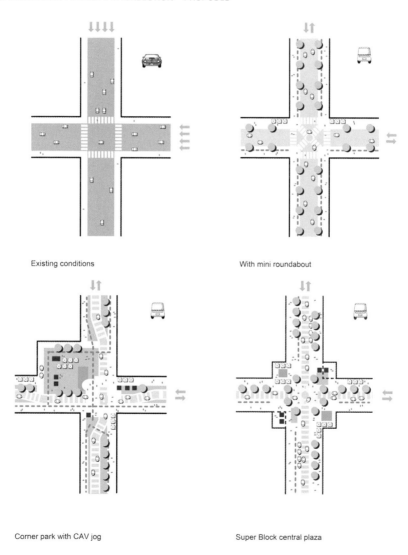

Existing conditions

With mini roundabout

Corner park with CAV jog

Super Block central plaza

85 per cent is reserved for automobile use. However, this is probably one of the best intersections in DTLA for pedestrian comfort and safety.

Figure 6.20 shows how the intersection could be reimagined following the CAVs supergrid strategy. The intersection would have a maximum speed limit of 15 mph, with two converging local movement streets. Shared CAVs services would use it only if the origin or destination of the trip is within the superblock. Less than 50 per cent of today's surface will be needed for CAVs movement, leaving the remaining square footage for other uses and users.

The Figure 6.20 diagrams illustrate how the intersection could be redesigned in combination with the transformation of an adjacent parking lot into one of the public parks described. These new public spaces should be designed as pedestrian-first type environments, and with a 'shared surface' approach. Other users would have to negotiate space among the different modes with minimum or no traffic separations.

Demarcation of users' space could be achieved by differentiating street surface materials and patterns, through landscaping, placing bollards strategically or by adding other street furniture elements. CAVs drop-off and pick-up space is provided along local movement streets kerbsides. The geometry of the street is redesigned to accommodate new functions with added active programmes within the street section. These new public, or publicly accessible, open spaces provide an opportunity to create a mix of uses and users and promote sociability, and will be programmed to be active for most of the day and part of the night.

In the redesign of local intersections, it is important to preserve minimum widths and areas for ecological functioning, and to increase ecological value. These spaces offer potential areas for the natural filtration of water into the ground, including run-off from hard surfaces from the intersection itself and from adjacent residential and commercial areas. Impervious surfaces should be limited and replaced with alternatives to improve infiltration and percolation.

PAVEMENT KERB

CAVs supergrid streets in urban cores should be created with the intention of becoming *living streets* – streets designed primarily with the interests of pedestrians in mind; social spaces where people can meet and with a shared space approach that will greatly reduce the demarcations between vehicle traffic and pedestrians. When all vehicles are CAVs, pavement kerbs as we experience them today would disappear, as there will be no need to discourage drivers from parking or driving on pavements. Kerbs in some conditions could still be used to channel run-off water into storm drains and to keep pavements areas dryer, or could be retained for aesthetic reasons.

Most streets in the CAVs supergrid model should be redesigned at the same grade as pavements, without kerbs, or with minimum kerb heights (1–2 inches) to allow drainage. This allows maximum pedestrian comfort and accessibility for people with limited mobility, while CAVs should be limited to a speed that does not disrupt other uses of the streets. Demarcation of the space of different users is achieved by a rich physical environment of contrasts in terms of surface tactility, materials, patterns, colours and the enhancement of sound and other sensory clues, and also by using bollards and landscape elements. Minimum kerb heights could be used to help visually impaired people to navigate.

The need for drop-off and pick-up space would increase exponentially. In urban cores, this would mostly be provided along streets near kerbsides, in a public and shared type of environment, and would serve multiple parcels and different types of users through the day (and night). These areas, or HODOs, would have pre-set or real-time waiting limits regulated and charged by the local authority. For passengers, this could be free for up to 4 minutes during peak-demand times.

CAVS MOBILITY HUBS (NEMOHs)

Ideally, CAVs parking and waiting stations in urban cores will be provided by a system of hubs organised around a core and a district parking strategy regulated by local municipalities. These hubs, or NEMOHs, would not only accommodate CAVs and their needs – waiting, parking, recharging, and servicing – but also serve the users through their urban journey, providing them with a menu of mobility options, services and shops, goods delivery and pick-up options, and a comfortable place to wait, relax and socialise. NEMOHs of different types and sizes serve different but correlated purposes. The CAVs supergrid strategy identifies

FIGURE 6.21
AN EXAMPLE OF A CAVS MOBILITY HUB

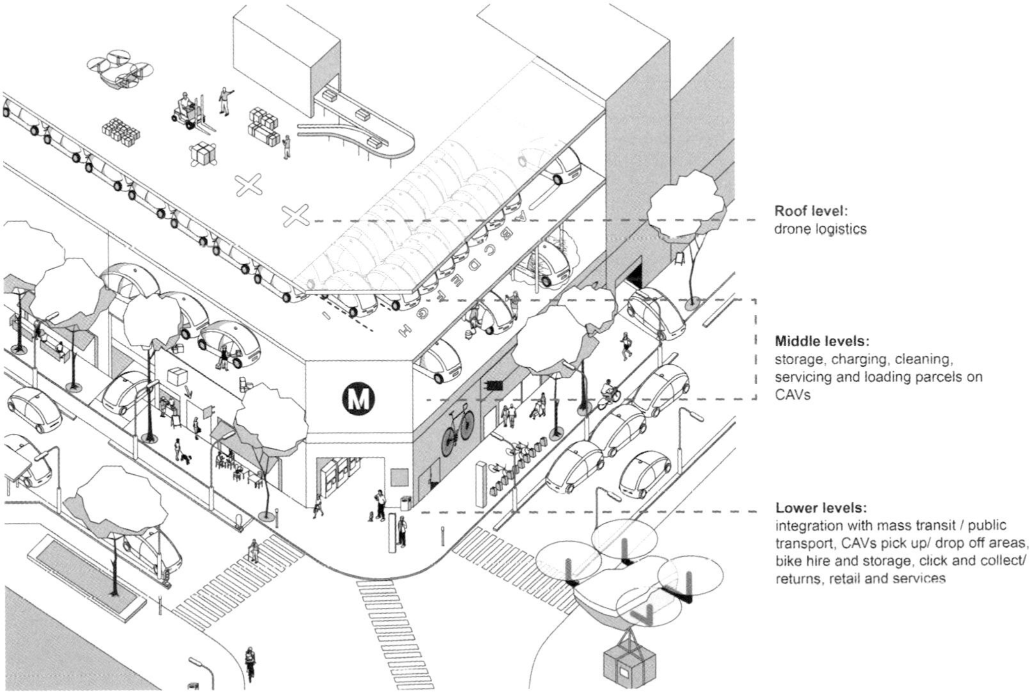

Roof level:
drone logistics

Middle levels:
storage, charging, cleaning, servicing and loading parcels on CAVs

Lower levels:
integration with mass transit / public transport, CAVs pick up/ drop off areas, bike hire and storage, click and collect/ returns, retail and services

a four-level hierarchy of CAVs hub: *core mobility hub, district mobility hub, superblock mobility hub, and remote mobility hub*.

Real-time congestion charges in combination with charging per mile enables cities to regulate the amount of privately-owned CAVs reaching cores and also where they will park. Privately-owned CAVs will drop off and pick up owners in a dedicated hub at the periphery of the core and then self-park in a remote hub in a less valuable location as directed by congestion charge incentives or disincentives and depending on time of day and waiting time. From these core hubs, passengers will be required to switch for their last mile to a shared CAVs system or to use public transport, cycle, or walk to reach their final destination in the core.

Multiple district mobility hubs located along entry points provide a more capillary-based system to serve individual districts within the core. Smaller superblock hubs provide fine-grain accessibility and services at the superblock level. In most cases, these types of hubs would be located along district movement streets. Real-time congestion charges – depending on factors such as route congestion, nearby parking availability and waiting times – will regulate shared CAVs parking distribution within the different facilities of the district.

PARKING AND OTHER IMPLICATIONS FOR THE BUILT FORM

In a CAVs supergrid scenario, parking policies and the way in which vehicles access properties change radically. These factors trigger a cascade effect in real estate regarding the design and layout of buildings. In an area like DTLA that has been dominated for decades by automobile-driven design choices and parking infrastructure, this entails a number of parameter changes that bring a radical rethinking of building typologies.

Minimum parking requirements
Current requirements for minimum parking standards should be replaced by maximum CAVs parking standards. Typically, this entails buildings in the future not being allowed to accommodate private CAVs parking. Few spaces, if any, will be provided to self-park shared and servicing type CAVs.

Maximum CAVs parking requirements
Maximum CAVs parking requirements for buildings include the maximum amount of kerb allowed for drop-off, or the extent of HODOs, on public streets and will regulate the maximum utilisation allowed for a shared HODO.

Kerbs and HODOs
Typically HODOs and kerbs used for CAVs drop-off, pick-up and service, are shared among multiple properties within walking distance, and are regulated by the municipality. HODOs could be taxed by use to recover some of the municipal revenue losses from parking revenue and tickets.

On site drop-off areas
On site drop-off areas should generally banned for private properties or restricted to few specific uses.

Building frontages

As on-site parking and parking lots disappear or shrink substantially, buildings will increasingly have less 'back-of-house' space and more frontage. This is particularly relevant in Los Angeles, where most building access today is located at the rear or from wherever car parking is located, creating a condition where the 'back of house' is the dominant feature.

Transitional space

The relationship between the public space of the street and the private sections of buildings would improve with the increased importance of lobbies and transition areas. In fact, these will be the prime connectors to and from buildings to HODOs for shared CAVs. A process of cross-contamination of functions and spaces will promote the creation of diffused semi-public space at the ground floor and on public pavements.

Flexibility buildings

Can be designed for flexibility to account for changes in parking regulations. The structure of the building, including the design of the elevation and positioning of openings, should be designed to accommodate future changes in use. On-site parking structures can be minimised and designed for future adaptation and easy conversion to habitable spaces. The minimum floor-to-floor clearance of parking structures should be 11 feet (3.5 metres), and should adopt flat floorplates or stacker systems that allow for future retrofit.

INTERIM PHASE

The CAVs supergrid model is not a futuristic vision, but a strategy that the Los Angeles region could start implementing in its cores incrementally from today. The next 20 years will be an interim phase in which CAVs and manually-driven automobiles will coexist in the majority of Los Angeles' streets.

The opportunities created by shared CAVs should be looked at closely, as they could be a strategic tool to solve the current mobility issues of Los Angeles' cores. CAVs will be able to make a difference only if deployed on a large scale. In this sense accelerating adoption will be critical and local authorities in DTLA will be required to take drastic decisions. They will need to redistribute space between different users by allocating dedicated corridors to shared CAVs from which manually-operated cars are banned. This is to ensure that there will be minimum mixing of the two types of technologies. They will also need to speed up the creation of a significantly improved cycle infrastructure. In the short term cyclists will be the most disrupting element for CAVs, therefore a world-class cycle infrastructure will limit conflict with CAVs as well as responding to rising demand.

C-Corridors should complement the traditional hierarchy of the road network but may not necessarily be part of it. In the beginning, corridors could take the form of a separate layer that will meet the road network used by manually-operated cars and public transportation only at key points. The diagram in Figure 6.22 suggests how during phase one of the interim phase corridors with shared CAVs and public-transport-only lanes could be progressively introduced in DTLA.

Phase one should jumpstart the transition by creating a loop of dedicated shared CAVs and public transport corridors, through downtown from Union Station in the north to Exposition Park in the south. This will connect most of the rail stops, landmarks and parking areas from the Civic Centre district to the University of Southern California (USC) campus to serve the majority

of commercial and residential areas for local trips and as a first/last mile alternative. In this scheme, Figueroa and Main Street could accommodate the main north-south CAVs movement C-Corridors, and be the first to be reconfigured as multi-way boulevards with dedicated shared CAVs and public transport lanes, along with the two east-west connections.

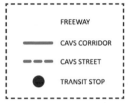

FIGURE 6.22
PHASE ONE CAV SUPERGRID

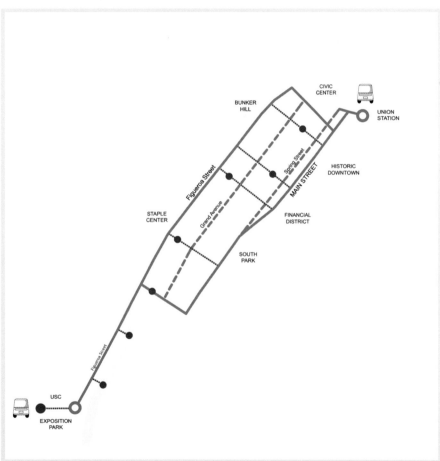

Secondly, within the loop a further CAVs hierarchy can be introduced for local streets. Grand Avenue, Spring Street, and Broadway Street, because of their unique civic, mixed-use or historical characters and identities, should be among the first sites for street reconfigurations and CAV-only services to be offered. As CAVs technology improves and becomes more dispersed and accepted, the CAVs supergrid can be progressively delineated within the loop area and mobility hubs can be established around the perimeter for easy transport mode exchange. Manually-driven cars at this stage will be progressively discouraged to access the area within the loop.

Once phase one is consolidated for the central part of the core, the same approach of C-Corridors can be expanded east of Main Street to reach the Fashion and Arts Districts, and ultimately the LA River and north to incorporate Chinatown. A dedicated shared CAVs and public transport corridor could work, for example, on Main Street. It will have a similar section to a district movement street, with the exception that traffic direction could be one-way or two-way and manually-operated cars could use part of the section. Central lanes will be reserved initially only for private and shared CAVs and public transport, with private CAVs excluded in a second phase. Side lanes could still be accessible

to manually-operated automobile traffic. Speed limits will be similar to a district movement street. The intent of phase one is to progressively increase accessibility and extension of the CAVs system and to gain acceptance from the public by giving CAV-type services, microtransit and – importantly – regular public transport priority over private automobiles. Phase one will also be the time to find the political will to start reclaiming space from car use to better use.

Many of the CAVs supergrid ideas and concepts discussed will already be applicable during the interim phase, as the different levels of CAVs automation will be able to navigate street environments as we experience them today while coexisting with cars.[16] Furthermore, these concepts are applicable even if the CAVs strategy is not applied in full, and instead an efficient public or private multimodal system is developed around the same framework. One of the ultimate goals is, after all, to drastically reduce the need for single-occupancy vehicles to access and move through the urban cores of Los Angeles. ▬▬

Chapter 7

A WAY FORWARD

Automobiles are often conveniently tagged as the villains responsible for the ills of cities and the disappointments and futilities of city planning. But the destructive effect of automobiles is much less a cause than a symptom of our incompetence at city building.

Jane Jacobs, *The Death and Life of Great American Cities*

The change and reconfiguration that the CAVs revolution will bring to our urban environments is significant. This is only the beginning, as both the technology and its interpretation by design and planning communities will evolve over time. The strategy presented and its application to areas of London and Los Angeles is designed to showcase the potential to plan for CAVs in a way that facilitates positive reconfiguration of mobility, public realm and buildings in urban cores. Furthermore, the strategy suggests how CAVs could trigger a larger process of urban regeneration and economic development, ultimately reshaping architecture by responding to the needs of new lifestyles. Most importantly, it suggests how CAVs could be used as a tool to improve the quality and performance of our cities and to address some of the most intractable issues they face.

A softer version of the approach has been presented for the deployment of CAV services for London. With the right political will, this strategy could be implemented incrementally from today, and could be adapted and refined to be applicable to many other European city cores with similar pre-industrial urban configurations, such as constrained rights of way, high densities, diverse mixes of use, and multiple layers of history, heritage, culture and governance. This implementation model would help in introducing CAVs while also unlocking potential for walking, cycling and environmental improvements. It also showcases the benefits of CAVs in terms of reduction in private car usage in urban areas.

In Los Angeles, by contrast, a more radical implementation of the model is proposed. This focuses around the urban grid plan, probably the most distinct and democratic characteristic of an American city, and typical of colonial and post-industrial development patterns. In this case, the discussion starts from the long-term framework of the CAVs supergrid and proposes a broader vision with a defined first phase that could be implemented relatively soon and would be easily adaptable to other cores. American city centres benefit from having space available for transformation and possible reuse, thanks to the original dimension of the grid plans and to more generous standards for development than in European cities. But American cities must also overcome decades of persistent car-centric urbanism and the rooted culture created by it – rights of way, for example, are enormous compared to European examples and typically cores display less mixing of use and fewer layers of historical stratification.

The comparison between London and Los Angeles illustrates how transformative a CAV-based strategy can be in very different settings and markets, with different societal values and lifestyles. In both cases, it is evident that a contextual approach is needed. Even if these two examples of approach are at opposite ends of the built environment spectrum, both would still have to overcome very similar hurdles in terms of public acceptance. This is why CAVs approaches such as these should be tested as soon as possible by undertaking regulated trials of driverless cars on the streets. Just as CAVs learn by driving, cities can learn to adapt to driverless cars and can exploit them to solve their issues.

CAVS AND THE URBAN ENVIRONMENT

CAVs are connected electric shared autonomous vehicles, which is a specific type of driverless car. This model was purposely selected for its potential to give something back to city dwellers rather than taking away their quality of life. The main proposition made here is regarding the 'shared' element, as at the moment there is no indication that the car of the future will be a shared one. To create a positive framework for street design, tackle environmental matters and manage rising population levels, it is necessary in city cores for CAVs to be shared. A Shared CAVs concept firmly contrasts with a worst-case scenario of amplifying current congestion patterns at unsustainable levels, with more people moving to urban areas and therefore more potential CAVs users (including teenagers, the elderly and the disabled) and ultimately more private mobility. This is a dangerous mix with the potential to bring cities to a halt. For this reason, CAVs have a real chance to express their benefits only if shared.

Furthermore, to capitalise on the potential benefits, CAVs will have to combine a fundamental set of characteristics. In particular, they will need to be: *electric* – to cut down pollution in urban environments; *connected* – to optimise and make more efficient use of the road network and free space;[1] *shared* – expanding the number of passengers per vehicle[2] and expanding the pool of users to elderly, teenaged and disabled users, while contributing to freeing more space for other uses; and *flexible* – through public and private partnerships CAVs services could help make fixed public transport service more functional and accessible, particularly by filling first/last mile gaps.

The other characteristic elements of CAVs, that is connectivity and automation, are the essential technological ingredients that have converged to make driverless cars possible. The technological quantum leap has been prepared by a convergence of different innovation, however the availability of technology alone is rarely sufficient to ignite change outside the research labs, unless there is a strong economic bait that unlocks investment – in this case the promise of making a fortune by getting a slice of the profitable automotive market. The sector is evolving fast and the boundaries between car manufacturers, IT firms, fleet operators and many others is rapidly blurring, with both new and traditional players flocking to secure a chance in a market that until very recently was a closed shop of automobile makers protected by formidable entry barriers. Things seem to have changed and, at the moment, no single interest has all the credentials to emerge as the dominant player, hence the enormous amount of acquisitions, alliances and collaborations taking place. A galaxy of players has been at work for a few years, spending billions in CAV-related technology and they are extremely committed to a return on their massive investment. This, in the end, is the reason why CAVs will probably become a reality on our streets, rather than any expected benefits of this new form of transport.

Over the course of a century, cars have profoundly changed urban environments in terms of dimensions, structure, form and quality. CAVs promise further developments in all these areas:

Dimensions

If driverless cars are shared in urban cores, they will offer space savings on streets and plots, which can result in better use of precious land and in denser neighbourhoods. However, CAVs also have the potential to prompt more sprawl, therefore they should be applied to suburbia with particular attention, in a way that promotes densification and full integration with public transport systems.

Structure

If CAVs are developed to trigger the retrofitting of the existing automobile infrastructure, cities won't see the dramatic levels of spatial reorganisation previously witnessed with cars. In the transition phase, however, there will be processes of redefining neighbourhood boundaries and city focal points, and addressing the potential gentrification prompted by the implementation of CAVs corridors and superblocks.

Form

New building types will emerge and their exact nature will be determined by the evolution of vehicles and, more importantly, by the changing lifestyles of city dwellers in response to the new technology. New architectural opportunities lie ahead in relation to mobility hubs built around CAVs, but also for neighbourhood logistic hubs covering the last mile. A trend towards more publicly-facing buildings is anticipated, as on-plot parking will disappear and pedestrian interfaces with the street and kerb will increase in importance.

Quality

If space gains are secured for the purposes of greening the city, the impact of CAVs will be revolutionary and will make a substantial difference to the sustainability and amenity credentials of urban areas. CAVs will provide the opportunity to turn roads back into streets, recalibrating their roles as socio-economic platforms, decluttering them from traffic paraphernalia, and reducing the uniformity effect to bring local character back to the foreground.

CAVs are not the only innovations on the horizon. Fuelled by technology and start-up investment, innovations ranging from dockless shared bicycles to shared electric scooters and MaaS are already a visible presence in our cities while many other projects are at prototyping phase, including straddling buses, or the Hyperloop. In many instances excitement will fade as quickly as it emerges, and only a small proportion of these novelties will in the end be realised to help city dwellers rather than just investors. However, some of these innovations could be relevant to the CAVs strategy. Current microtransit experiments offer some interesting lessons, despite a number of limitations and dangers, including the potential to erode the role of public transport. An example is City Mapper's Smart Ride in London[3] where a company has designed a network, rather than a route, for a fleet of ride-hailing vehicles so that the service has the flexibility to adapt the routes to demand. The concept of 'bus stops' therefore evolves, and passengers are picked up at a variety of locations via an app. The network is allowed to change and adapt within a certain urban area and, as current regulations do not allow vehicles delivering this type of service to be buses, the company is using vans, claiming to be 'vehicle agnostic' and thus opening the door to a number of potential developments. This existing model could very well work with driverless cars. Regardless of ultimate destiny the implementation hiccups of this type of service have delivered some important lessons.[4] The interesting element in terms of design is the concept of a responsive network within defined urban areas, which is compatible with, and can even be seen as a precursor to, the proposed approach of CAVs Corridors, cells and superblocks. It also shows how creating a synergy between private and public sector initiatives will prompt the basic ecosystem to facilitate the introduction of CAVs. In this direction in London, some local authorities are experimenting with simple yet innovative schemes looking at the feasibility of removing general traffic from sections of busy central corridor to favour cyclists and buses. By chance, one of these initiatives is in Shoreditch where the initiative City Fringe Ultra Low Emission Streets is banning all vehicles other than electric, hydrogen-powered and a limited number of hybrid vehicles from a portion of the neighbourhood in mornings and late afternoons.

IMPLEMENTATION AND RECOMMENDATIONS FOR CITY MAKERS

The major challenge for city makers in introducing CAVs is to positively guide the transformation of urban environments, fostering public acceptance and uptake of this new technology while progressively discouraging single-occupancy private automobile use. It is important to get this transition right. The dangers lie in creating another difficult roads legacy that will work against the principles

discussed in this book. This risk is real, and quick adoption will be important to securing the benefits of CAVs. Big choices and investment will be required for instance in banning manually-operated cars and, successively, private CAVs and in creating dedicated C-Corridors or mobility hubs.

There are also many 'small' choices that could create challenges, for example in terms of dimensions. Allowing CAVs to follow the car growth trend that has led to many vehicles now being more than 2 metres (6.5 feet) wide would result in losing some of the gains from efficiencies in space allocation. Allowing excessive platooning will affect street vitality, and not managing and sharing space between servicing, deliveries, and CAVs drop-off will radically reduce the amount of space available for environmental improvements, as would not limiting the sections of drop-off areas along pavements and so on.

Finally, a clumsy adoption of CAVs in places where walking, cycling and the use of public transport is already the most natural way to move around the city may also lead to a decline of sustainable habits in favour of the use of CAVs, in a way not dissimilar to that of pedestrians switching to e-scooters.

The ultimate goal is to promote a pragmatic, human-centred vision for CAVs and the city within which different interests and the public could converge. Below is a set of preliminary recommendations for city planners on general policies regarding the introduction of CAVs to urban cores.

People then technology
Don't delay people-centric CAVs trials, and work with manufacturers to draw lessons – plan from this point forwards.

Show off
Use demonstration projects to test and showcase the potential of CAVs. Not only CAVs testbeds but real-life pilots are needed, which focus on how CAVs can improve cities and people's lives. These could be corridors or entire city quarters – as recently explored in Barcelona for the superblock. Change won't happen overnight but pressure will mount soon. Why not accelerate the erosion of car space today, as Copenhagen has done? [5]

Power to the people
Space gains should be redistributed according to the movement hierarchy principles set out in this book – pedestrians first followed by bicycles and public transport, then CAVs and lastly cars. They should also take into account the introduction of blue and green infrastructure.

Hail to the filling station
Safeguard some of the soon-to-be redundant car infrastructure for possible reuse by CAVs. Once this infrastructure is lost to other uses, such as offices, it will be much more expensive to find space in what are often strategic locations.

Death to the parking ticket
Plan for a change in revenue sources for local authorities, and start experimenting with different ways of generating income from use of the road network such as kerb utilisation, miles travelled, or congestion charges.

Density matters
If parking revenues are soon to be out of the picture, many sites may become newly financially viable or physically accessible. Use this opportunity to build denser neighbourhoods that optimise the use of land in areas already served by infrastructure. This should include switching from minimum to maximum parking requirements.

Smart cities light
Technology is still evolving. Local authorities should consider carefully before they commit to expensive infrastructure, and should capitalise on the ability of CAVs to take complexity on board by designing smart low-tech streets instead. Resources may be better spent in finding ways to coordinate CAVs movements, especially if the fleets have different owners.

People and goods
Explore synergies between personal mobility and logistics, particularly for the last mile of deliveries. Can facilities be shared? Can they use the same CAVs?

Suburb 2.0
Start planning for the integration of CAVs and public transport in existing suburbs. Suburbs can be better connected by using CAVs to solve their intrinsic structural problem – the first/last mile connection to public transport. Existing suburbs can be densified to absorb a more substantial share of population growth in a sustainable way.

Design for flexibility
Promote flexibility in building design, particularly for parking structures. Buildings today will need to be designed for flexibility to account for tomorrow's changes in parking regulations. On-site parking structures should be minimised and designed for future adaptation such as easy conversion to habitable spaces – flat floorplates and generous floor-to-floor heights will allow for future adaptation.

The public sector as well as developers and asset managers will need to catch up with the incredible challenges and opportunities created by CAVs. It is concerning that in Europe little or no consideration has been given to CAVs in recent policy documents that are meant to guide the planning and development of urban areas over the next two decades.[6] In the US there are some attempts by regional and local authorities to look ahead but typically these are generic demands to take into account future technological advancements rather than setting the agenda for how CAVs should deliver social benefits.[7] Meanwhile, urbanists generally have been disengaged from the topic with the risk of leaving the planning field open to technologists, manufacturers, or other specialists who are not necessarily equipped for, or interested in, setting the best trajectory for social well-being or urban sustainability and competitiveness. In the UK the population will reach 73 million by the beginning of the 2040s.[8] The United States will witness a population increase of over 100 million people in the next 50 years.[9] Similar or more pronounced population growth patterns are predicted for the rest of the world.

At the same time, urban development globally remains rooted in a heavy dependence on the automobile. Population growth forecasts in conjunction with climate change challenges mean that formulating a pragmatic vision for CAVs and urban design is more relevant than ever. CAVs could be a sustainable answer to urban transportation, opening up a post-car world paradigm and creating opportunities for resilience and climate change adaptation thanks to the changes they could potentially bring to the fabric of cities. Achieving successful results without a strong CAVs strategy will entail an enormous investment in public transport even where densities of catchment areas would not necessarily justify it. Societies are likely to eventually accept the concept of being driven around by automatic cars, but are we also prepared to surrender the control of our cities and miss the unique opportunities at hand?

As urban designers we look forward to the opportunities that CAVs will bring to improve our urban environments. In continuing our work, we will test these ideas on other city fabric types and against different contexts, including other world cities with different characteristics. Our ultimate goal is to provide a positive vision for CAVs and the city, and to elaborate a coherent urban design framework to provide rational plans for our cities when CAVs will be the new normal. It is hoped that this work will contribute to the formulation of a pragmatic vision through which different interests and the public can converge. The imperative for urban designers must be to explore with creativity and imaginative skill the opportunities at hand and to provide – with a mix of authority and deference, flâneur-like saunter and indefatigable optimism – a vision for the renaissance of the human-centric city in the age of the driverless car. ■

2.10   2.40
6.89   7.87

APPENDICES

APPENDIX I

CAVs and urban design

Mike Davies,
Architect and Urbanist,
Rogers Stirk Harbour + Partners

Technological change can be uncomfortably rapid and can force new thinking. When our firm was designing the new building for Lloyd's, in the City of London, I had to interview the 300 syndicates about their future needs. At the time, only one underwriter out of nearly 3,000 had any form of advanced technology at their desk – a small cathode-ray tube monitor on which green dot matrix sentences would glow.

On asking the syndicates about their IT computing needs, the message I received was: What is IT and what are computers? They would certainly not be necessary. The activities are broadly the same today, but data that used to take days and sometimes weeks to amass from around the world became available by fax and computer terminals. By the time the building was finished, many syndicates were on their second generation of computer. A speed and data revolution had swept in – with once doubting syndicates responding to change and opportunity rapidly – bringing huge positive benefits to Lloyd's and keeping them at the forefront of the world-wide insurance market. The impact of the next revolution, the internet, was still some years away.

Autonomous Vehicles

Evolution of transportation technology has always been directly linked to city life and its urban and social fabric. Our cities will be transformed positively by the emergence of multi-tasking autonomous driverless vehicles incorporating network route planning, drop and pick-up on demand and locational and traffic awareness. Fleet-owned, leased and private autonomous vehicles will become the lifeblood of the city street, alongside automated buses.

Automated electric vehicles are virtually pollution free – with pollution control being at the power station or with negligible pollution, if from clean generation sources such as wind and solar farms. Citizens will be safer. Road deaths will diminish and cyclists, pedestrians and other road users will use the city in greater safety alongside autonomous vehicles. Compact autonomous vehicles take up only two thirds of the parking space of traditional cars. Combined with a drop in the number of traditional vehicles entering the central city zones, up to 50 per cent of city parking area can be transformed back into public space as tree avenues, planting and public amenities.

With big reductions in traffic noise and pollution, the quality of pavement life in city streets will improve dramatically. Café-terrace life will expand. Markets, flower sellers, transient public events and amenities will flourish. Autonomous vehicles charging and dwelling on the electrical power grid represent a huge energy reservoir – which can buffer demand peaks in national and local energy networks, working in coordination with power stations and intermittent-supply solar and wind farms. About a third of modern city traffic is goods and services delivery related. Autonomous vehicle timed deliveries will displace traditional delivery methods. Drone carrier vehicles will provide for the last few hundred metres to the delivery bay and others will act as mobile multiple drone delivery hives.

Ease of movement within a city is a measure of its quality of life. Autonomous vehicles will act in coordination with the city traffic flow control networks and controllers, cooperating to maximise ease of local flows and longer distance movement across the city. Autonomous vehicles will provide chauffeured transportation in areas with poor local services, bringing citizens to otherwise inaccessible suburban transport nodes and supporting those who wish to engage with the city.

Local suburban rail/underground/tram transport nodes will densify, reinforced by autonomous vehicles bringing and returning dwellers too distant from the station to walk – replacing private car use to the station or directly to the city heartlands – offering a green personalised local transport service, either individually or group owned.

Autonomous vehicles will provide excellent service to the elderly, the infirm, the school run and to those who choose not to or cannot drive. The social value and convenience of these silent helpers will add to the quality of life for many special groups of citizens.

Will we be unprepared for the transition?

Local traffic management, councils and planners will need to explore the implications of personalised light transport. Public debate, movement and operational studies, autonomous vehicle law, ownership structures, new opportunities and community studies, AV travel club structures and test scenarios should all be rigorously explored.

The automated vehicle is not simply a technological development but also an ecologically responsible social, community and citywide tool. There is little doubt that the convenient, safe, self-driving autonomous city car, available to all, will change our cities and our urban habits in positive ways, offering opportunities and individual and community support hitherto unimagined.

European evolutionary cities will benefit from the introduction of highly adaptable smart autonomous vehicles, which will, through evolution of use patterns and services, become an indigenous part of the city and its incremental growth. Revolutionary technology will integrate easily within old growth, urban centralised city fabrics.

At the other extreme Los Angeles, a grid city, is set for true revolution. The arrival of the smart autonomous vehicle will cause seismic change in a city where a third of the urban surface is dedicated to the car. Giant parking lots all over the city will become redundant, as the autonomous vehicles work much of the time and only sleep when charging. Parking requirements in LA will shrink dramatically, leaving vast tracts of parking area free for development – a land supply glut.

As a result, Los Angeles will respond by seizing the opportunity to create new public parks, urban forests, open spaces, and urban agricultural and ecological areas within the city – opportunities

unimaginable previously – taking big and surprising steps towards a better ecologically balanced city, and a response completely unlike that of European old-world cities.

The autonomous vehicle of Los Angeles will be a cruising vehicle capable of long-distance runs in a world of freeways and huge urban distances, quite unlike the small urban runabouts that will appear in the grain of old cities.

LA is nearly 100 miles long by 35 miles wide, with only an occasional enlightened hint of public transport – in Los Angeles the car is still the king, and the autonomous car of LA will adapt to the city as well as the city adapting to the autonomous car. In each world city, the autonomous car will play a different role, adapting to the opportunities and use habits of each city.

What is certain is that the autonomous vehicle, the rapidly approaching next step in urban transportation, will have profound but different impacts on each different city. It will fit in incrementally, adapting its service and habits to different context.

The power of the autonomous vehicle is its infrastructure-independent adaptability and its individual and collective brainpower.

As such, the autonomous electric vehicle will spread everywhere, as convenience is the wish of every citizen in cities worldwide.

FIGURE AI.1
LLOYD'S BUILDING, LONDON,
CONSTRUCTED 1978–86

APPENDIX II

An interview with Andres Sevtsuk

Assistant Professor of Planning
at Harvard Graduate School of Design
in the Department of Urban Planning and
Design, and Director of City Form Lab

As a practising urban designer and educator, what impacts do you think this new technology will have, if any, on city making in North America?

If history is a guide, then automated vehicle technology is likely to become a major influence on city form. Driverless vehicle technology is introducing significant cost savings to both individual as well as fleet-based vehicle owners. While moving a passenger with a human driver currently costs transport network companies (TNCs) like Uber and Lyft around two dollars per mile, automated vehicles (AVs) are projected to initially slash these costs to less than a dollar per mile – on par with the total costs of owning a personal car. Eventually, as technology improves and adoption increases, the cost of autonomous robo-taxis is expected to decrease to around 20 cents per mile – lower than the current gasoline and parking costs of a private car.

Such a drastic cost reduction will lead to a massive increase in vehicle demand and vehicle miles travelled. Even if automated vehicles initially run as on-demand fleets, reducing the number of vehicles needed to move the same number of passengers, traffic on streets will significantly increase. Whereas conventional human-driven cars stay parked for around 95 per cent of the time, the combination of slashed costs and new fleet-based business models will ensure that robo-taxis are used far more consistently than personal cars. Paradoxically, a city with fewer cars will have far more trafficked roads. Freeing the car occupier from the inconvenience of driving will also increase tolerance for travel times and distances, leading to a renewed push for low-density suburban growth on the urban edges, and car-oriented lifestyles.

Widespread ride sharing is often touted as the potential saviour of American cities from a second wave of sprawl and automotive infrastructure that AVs could spur. Yet current data suggests that these ride sharing expectations could be grossly exaggerated. If market forces get their way, automated vehicles in North American cities will provide a strong push against much of what environmentally and socially conscious urbanists have advocated for in the past decades. They will further spread out land values, increase sprawl, reduce the density and diversity of existing urban centres and challenge public transit as well as active transport mode shares. They will also reduce pedestrian and bike priority in dense and mixed-use urban environments, instead enforcing both legal and economic advantages of individual cars on the public right of way.

That said, whether and how AV technology will reshape streets, districts, cities and even urban regions depends a great deal on how municipal, state and the federal governments choose to intervene in the deployment of this technology. Many of the negative pathways can be turned into positive opportunities through forward-looking policies and regulations.

Could you describe in what way your research at Harvard is looking into the relationship between driverless cars and urban form?

The Future of Street research project at the Harvard Graduate School of Design investigates how cities might adapt streets to newly emerging shared, electric, and autonomous transportation technology in ways that maximise multimodal, socially inclusive, and environmentally sustainable outcomes. We are essentially interested in what it will take in terms of urban design, policy making and regulations to ensure that AV technology will actually foster

and increase public transit and shared ridership, walking, cycling and transit-oriented development. Undertaken in collaboration with the City of Los Angeles, the City of Boston, and industry partners representing technology, transportation, and infrastructure sectors, it comprises empirical research, spatial analysis, predictive modelling, and design studios.

Could you elaborate on the AV 'heaven/hell' scenario you and your team have been developing?

Some of the key uncertainties in AV deployment are a) whether or not the vehicles will be predominantly gasoline or electric, b) whether vehicles will achieve Level 3, 4 or 5 automation, c) whether they will be privately owned or fleet-based, and even in the fleet scenario d) whether the vehicles will be predominantly used for single-person rides, or shared rides with other passengers. The potential development pathways these questions introduce are shown in the scenario tree in Figure AII.1 below. The top figure indicates where we are today, with most vehicles in private ownership, gasoline powered, and human driven. The two other scenarios beneath highlight the best (middle) and worst-case scenarios (bottom).

Most interestingly, this uncertainty tree demonstrates that AV technology introduces many more problematic scenarios from an urban design and planning perspective than positive scenarios for sustainable city development. For example, regardless of the level of automation, if AVs are predominantly used for single passenger transportation, we will witness a significant increase in traffic and congestion on city streets, which in turn will reintroduce the already familiar consequences of low-density and high-energy city growth, reduced emphasis on existing urban centres and higher demand

Today's automobility

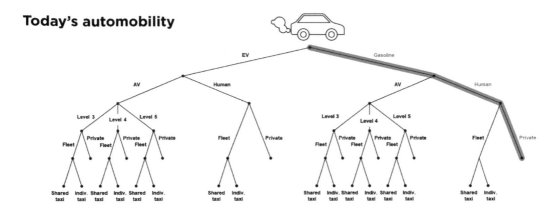

One positive scenario

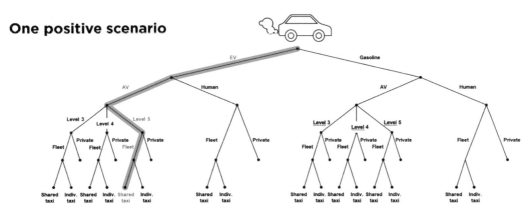

Twelve negative scenarios

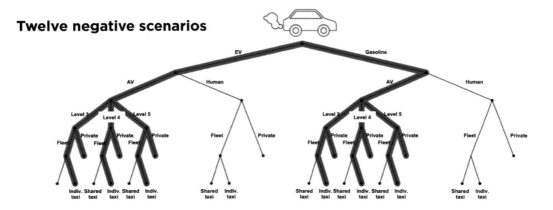

FIGURE AII.1
POSSIBLE PATHWAYS TOWARDS
UNCERTAIN AV FUTURES

for automotive infrastructure. Similarly, if AVs in the long run fall back on private ownership, we are also likely to witness more space and resources devoted to cars in the city.

There is, however, also one very positive future scenario – a potential future where AV technology is predominantly used for shared multi-passenger rides, essentially a reinvented public transit system. Vehicles are predominantly electric, Level-5 capable, fleet-based multi-passenger buses of various sizes. A city in this scenario will have a large fleet of fully autonomous electric buses ranging from neighbourhood minivans to full-fledged bus rapid transit (BRT) corridors running on intelligent vehicles that can easily navigate through the most complex multimodal streets full of pedestrians and bicycles. Most buses will still run on fixed routes instead of

on-demand custom paths, because this will guarantee predictability, efficiency and increased ridership. Heavy-capacity routes are surrounded by dense, mixed-use developments, enabling riders to conveniently walk to their nearest stations.

But unlike bus systems today, dropping the driver cost through automation will enable a significant expansion of routes, denser schedules and a lot more buses in the city. Municipal transportation departments will be able to maintain a far greater fleet of buses that serves every neighbourhood with the same budget as today. Only very low-density neighbourhoods and riders with special needs (e.g. elderly, pregnant, differently abled) will be regularly serviced on customised door-to-door routes. Others who wish to access individual AVs with door-to-door service will just have to pay significantly steeper fees, the same way as one pays for black cab rides in London today.

FIGURE AII.2
CURRENT BELOW, BEST-CASE (TOP RIGHT) AND WORST-CASE (BOTTOM RIGHT) SCENARIOS OF AV INFLUENCED INFRASTRUCTURE CHANGES AROUND VERMONT / SANTA MONICA METRO STATION IN LOS ANGELES, CA

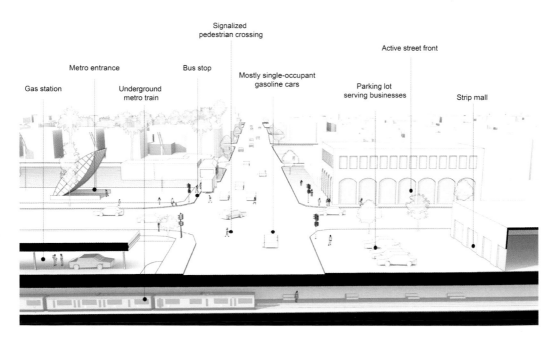

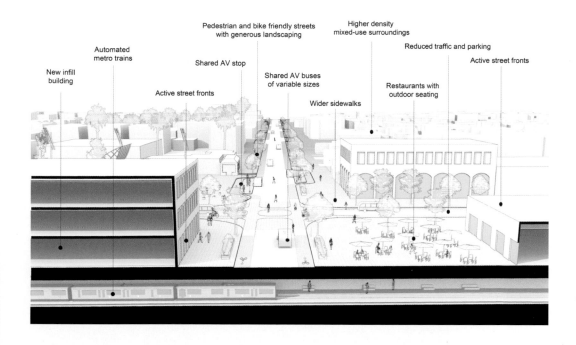

New infill
building

Automated
metro trains

Active street fronts

Shared AV stop

Pedestrian and bike friendly streets
with generous landscaping

Shared AV buses
of variable sizes

Wider sidewalks

Higher density
mixed-use surroundings

Reduced traffic and parking

Restaurants with
outdoor seating

Active street fronts

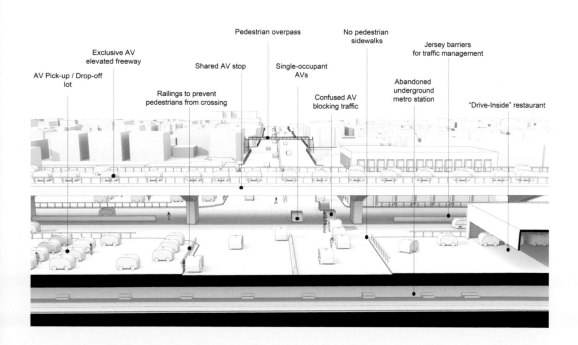

AV Pick-up / Drop-off
lot

Exclusive AV
elevated freeway

Railings to prevent
pedestrians from crossing

Shared AV stop

Pedestrian overpass

Single-occupant
AVs

Confused AV
blocking traffic

No pedestrian
sidewalks

Abandoned
underground
metro station

Jersey barriers
for traffic management

"Drive-Inside" restaurant

The drawings illustrate what these best- and worst-case futures could mean in the long term for a typical street in Los Angeles, CA. The first image describes the street today. We picked a transit-oriented location at the intersection of Vermont and Santa Monica that already has a heavy-rail Metro station to explore how AVs impact street design and urban form in a transit-oriented area.

In the best-case scenario, the number of individual vehicles is notably reduced on the street, replaced by buses of various sizes. Pick-ups and drop-offs happen primarily at designated stops and the roadway is reduced to accommodate wider pedestrian walkways and sidewalk activities. Streets are easy to cross on foot, and intersections are safe and give pedestrians priority. The reduced footprint of private vehicles and the increased reliance on shared transit leads to a lot more pedestrians on the street and opens ample land for densification, regeneration, landscaping and new mixed-use developments with active ground floors.

The worst-case scenario is essentially 'Carmageddon'. Priority on the public right-of way is shifted further towards AVs, most of which carry a single passenger at a time. This leads to increased traffic congestion, and pressures the city to widen lanes and to keep pedestrians, cyclists and human drivers off the roadway. Pedestrian crossings are far apart and lifted into elevated walkways. An exclusive, elevated AV highway is added above ground. The underground Metro line is left to decay, with ridership shattered by highly subsidised AV services on the roads. Pedestrians and ground floor activities on streets have been replaced by 'drive-in' businesses that accommodate electric AVs in building interiors. Some of the emptied-out buildings are converted into parking lots, AV maintenance grounds and temporary stalling areas.

In which area do you think that CAVs could contribute to solving urban issues?

From a spatial perspective, AVs are only marginally different from traditional cars – they take up a similar amount of space per person as a human-driven gasoline guzzler and they require ample land for storage and maintenance. Just like traditional cars, AV technology in individual cars exerts high demands on space and remains inefficient in terms of passenger throughput per unit area of land. This fundamental geometry constraint again suggests that the primary way in which AV technology can improve built environments compared to current North American settings is by deploying the technology primarily in shared, multi-passenger vehicles – by reinventing public transit. But transit ridership is low in the United States, extremely low compared to other developed countries around the world. Unless drastic measures are deployed to guide AV technology towards multi-passenger vehicles and public transport, it is more likely that positive urban impacts from AV technology will be harnessed in European and Asian cities that actively promote public transit and value efficient use of land much more than American cities.

One of this book's case studies is in the City of Los Angeles. What recommendations would you give to city managers to prepare for CAVs in a city where almost 90 per cent of daily trips are made by car?

My advice to LA would be to bank more on buses, while keeping up the long-term heavy rail investments and to keep convincing voters that new affordable housing allowances and higher development densities around station areas are bound to make the city more efficient, more liveable and more desirable

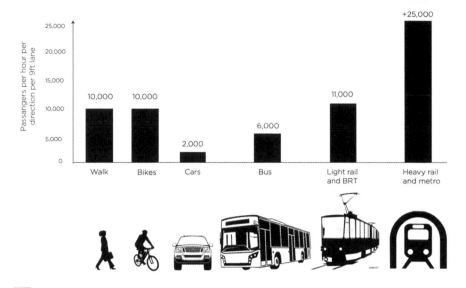

FIGURE AII.3
PASSENGER THROUGHPUT PER HOUR PER DIRECTION PER
9-FOOT LANE ON DIFFERENT TRAVEL MODES

in the long run. I think LA could also learn from Singapore, where the city is actively pushing AV technology into public transit instead of private cabs. Singapore is on track to open several AV bus lines in the next two years, setting an example to the world of how the public can benefit from this advanced transportation technology. I think committing to a fully automated and electric bus fleet by 2040 would help LA significantly reduce public transit operating costs, expand the network and improve schedules. A new, automated bus network could accommodate a range of vehicle sizes, appropriate for each context – smaller neighbourhood shuttles feeding busier routes, which in turn connect to high-capacity light and heavy rail. Implementing such a system would also give the city a unique opportunity to introduce 'transit-first' traffic policies, giving priority to buses and light rail cars at every intersection. Such policies are already in place in a number of transit-friendly cities like Zurich.

Cars and CAVs will have to coexist for a period of time as the old technology transitions out in favour of the new. What recommendation would you give to city managers and policy makers to positively drive this transition, also looking at lessons from the automobile era?

One of the potential hazards with AVs is that initial commercial deployments will involve Level 3 and 4 automation vehicles, which cannot safely roam in any complex urban environment and require special infrastructure provisions (e.g. clear lane markings, AV friendly intersections, few pedestrians on the road) to operate smoothly. I think every city deploying AVs should hold back making any special provisions for AVs on its streets and require that all AV services are capable of handling the most complex, multimodal street conditions before entering a city. Cities should also resist efforts from AV lobbyists to introduce any further restrictions on pedestrians on the roadway, and instead continue moving towards more pedestrian, bike and public transit friendly streets that organisations such as the National Association of City Transportation Officials have been promoting over the last decade.

Any other thoughts?

Cities would benefit most from AV technology if the technology is channelled into multi-passenger vehicles and public transit, instead of individual cars. Cities should also resist new infrastructure provisions for AVs, as well as other mobility innovations like package delivery robots, until the technology is mature enough to cope with the most complicated, crowded, multimodal street conditions like Times Square. And finally, cities, states and federal government should require that any accident involving an AV produces a fully transparent public record that is made accessible to all AV makers and operators so that one accident on a particular street could immediately enable millions of other vehicles around the country to learn from it and avoid similar accidents from that point onward. Such a policy has been in effect in the airline industry for decades. Yet in the aggressively competitive AV market, AV companies currently safeguard their data, holding back faster and safer system-wide improvements that would benefit the public.

ADOPTION

The adoption of CAVs will follow the classic stages of any other product or service and CAVs will need a number of years to penetrate the market. Accelerating this process, whether because of initial public diffidence or technical features to be perfected, may confine the use of CAVs to environments that offer specific characteristics in terms of predictability. This would limit risks and disruption, thus enhancing safety and performance. Spatial segregation from other modes of transport on city streets and the associated profoundly negative effect on city vitality would only be one step away from this, and represents a tempting quick fix with a difficult legacy.

ADVANCED TRAFFIC MANAGEMENT SYSTEMS (ATMS)

ATMS utilise intelligent infrastructure and real-time traffic data to improve traffic flow and vehicle safety. CAVs could revolutionise ATMS by enabling them to be informed by data inputs from every vehicle on the road, or ATMS could even control the movements of vehicles across the transportation system and maximise the efficiency of a city's roadway infrastructure in real time.

AIR QUALITY

Electric CAVs would immensely improve the situation in urban areas, not just by removing the emissions component from traditional cars but also by creating opportunities for urban greening, which would help to capture and neutralise air pollutants from other sources. Pollution will still be generated elsewhere, therefore in terms of environmental sustainability it is important to shift the focus from pure emissions control to demand management by promoting shared transport – another key area where CAVs can make a difference. Shared electric CAVs designed to maximise urban greening will be the next big thing for air quality and sustainability.

ARTIFICIAL INTELLIGENCE (AI)

The critical element of automation is AI, which consists of computer systems able to perform tasks normally requiring human intelligence such as visual perception, speech recognition, decision-making and, in the case of CAVs, driving.

AUTOMATION (LEVELS)

Use of machines that operate automatically. When applied to cars (autonomous vehicles), the level of automation varies according to the degree of automated technology embedded in the car. There are six levels of driving automation as defined by the Society of Automotive Engineers (SAE International). These are:

- Level 0: No automation. A human controls all driving tasks even when aided by warning systems.
- Level 1: Driver assistance. A human controls most driving; the vehicle performs either specific steering or acceleration and braking tasks.
- Level 2: Partial automation. The vehicle performs both specific steering and acceleration and braking tasks; a human controls all other driving.
- Level 3: Conditional automation. The vehicle performs driving in some modes; a human intervenes when requested.
- Level 4: High automation. The vehicle controls specific driving modes without human intervention.
- Level 5: Full automation. The vehicle controls all driving at all times without human intervention. See also autonomous vehicle.

While Levels 0–3 effectively represent business as usual in our cities, from Level 4 onwards car space starts to change as infrastructure is reconfigured and progressively scaled back.

CAPACITY

This is the maximum potential capacity of a road or a network, traditionally measured by vehicles per hour or day. This definition does not take into account other road users such as pedestrians, cyclists and public transport passengers. It is therefore not only inaccurate but potentially dangerous as it postulates that the way to improve capacity is by improving the performance of vehicles. Shared CAVs promise to improve capacity in a number of ways, including by reducing the amount of road space used per driverless car due to supposedly more accurate manoeuvring, by removing a large proportion of street parking, by platooning vehicles and decreasing the number of vehicles used via shared journeys, and by eliminating unnecessary circulation to seek out parking space.

C-CORRIDOR (CAV CORRIDORS)

A continuous sequence of streets designed to prioritise pedestrian, cycle and CAVs movement while banning traditional cars. In the transition phases when both cars and CAVs are on our city streets, C-Corridors could create a parallel network to car infrastructure, paving the way for their adoption.

COMPLETE STREETS

A set of policies and design principles from the US that aims to create safe, convenient, accessible and comfortable streets for all ages and abilities regardless of transportation modes. The objective is to rebalance the function of streets from the needs of the driver to those of all users, with benefits in areas spanning from health and local economy to safety and security. As with the UK's Healthy Streets approach, Complete Streets is a transport planning and engineering reformulation of urban design theory. CAVs can play an important role in delivering the outcomes sought by Complete Streets if street design enables increased accessibility and improved environmental conditions. Replacing cars with CAVs without changing the streetscape would contradict the ethos of Complete Streets, even if the design complies with all other current best practice.

CONGESTION

Overcrowding and blockage affecting the flow of vehicles and the overall performance of a transport network. CAVs are unlikely to be the solution to congestion and shared CAVs may not be a full solution, as access to this type of mobility is likely to increase ridership by those who currently have no access to vehicles such as children, the elderly or those with disabilities. Restrictions will need to apply to allow the network to move in ever denser cities. City design and planning will be the best way to mitigate the issue, by creating compact mixed-used neighbourhoods and by promoting sustainable forms of transport.

CONNECTED AUTONOMOUS VEHICLE (CAV)

A vehicle that integrates autonomous and connected technologies. A connected vehicle is equipped with technology enabling it to connect to devices within the car as well as with external networks such as the internet. Autonomous vehicles

are capable of fulfilling the operational functions of a traditional car without a human operator, typically classified as Level 3 automation or above. Both connectivity and autonomy create challenges for the designer due to the interactions between the vehicle and people and the built environment. The realisation of the benefits of CAVs will depend on the way CAVs are implemented and how the urban environment is designed to accommodate them – determining whether CAVs will lead to positive or negative future city environments. It is important to remember that CAVs are a means to an end.

CONNECTED VEHICLE (CV)

A technology that relies on information gathered by vehicles and the transportation infrastructure about real-time operations of the transportation network. A connected vehicle avoids traffic jams by finding the fastest alternative route in real time. CVs also inform other vehicles of their intended movements. In this way, vehicles could travel in a safer and more harmonious condition than today's traffic. CV technology is currently being tested in several cities and by several companies around the world.

CURB

(US) see **kerb** (UK).

ELECTRIC VEHICLE (EV)

A car that runs at least partially on electricity. Unlike conventional vehicles that use a gasoline or diesel-powered engine, electric cars and trucks use an electric motor powered by electricity from batteries or a fuel cell. EV would make filling stations redundant and spark the need for charging facilities. A number of

design issues would arise, from upgrading existing utility provision to introducing dedicated long-stay bays or integrating induction-charging routes and so on. Electric CAVs, by removing the emissions component from traditional cars, will drastically improve environmental conditions in urban areas.

FIRST AND LAST MILE

The first or last leg of a multimodal trip, typically from home or work to a public transport hub. Promoting the first and last mile means improving the convenience and efficiency of public transport. CAVs could play a big role in catering for first and the last mile journeys, however they are also a threat that could lead to discarding more sustainable travel options such as public transport, walking and cycling, ultimately resulting in a drop in physical activity and a Wall-E scenario of overweight citizens carried around by machines. Cities will still need to be designed to make walking and cycling the obvious first choice.

GEOFENCING

Feature in a software program that uses the global positioning system (GPS) or radio frequency identification (RFID) to define geographical boundaries. It has multiple applications for CAVs, including the definition of areas permanently or temporarily accessible by vehicles. They offer a light-touch alternative to conspicuous and often unattractive physical barriers in the public realm.

GROWTH MANAGEMENT

Specific regulatory policies aimed at influencing how growth occurs, mainly within a locality. These affect density, availability of land, mixtures of uses, and timing of development. Growth management seeks to accommodate growth rationally and is an effective strategy to combat sprawl and support transit oriented development. Since CAVs will open the potential for even longer commutes that will extend across jurisdictional boundaries, growth management strategies will be increasingly necessary across jurisdictional lines to constrain development within each jurisdiction to planned centralities.

HEALTHY STREETS (UK)

The Healthy Streets Approach is a system of policies and strategies developed by Lucy Saunders and adopted by TfL to 'deliver a healthier, more inclusive city where people choose to walk, cycle and use public transport' (Greater London Authority). It identifies 10 indicators to help measure performance and identify areas for interventions, focusing on issues such as pedestrian safety, the amenity of streets and combatting noise and air pollution. See also observation for Complete Streets and CAVs.

HODO (HOP-ON/DROP-OFF)

A specific CAVs infrastructure designed for the collection or delivery of goods and people with CAVs. HODOs would occupy dedicated bays, with vehicles allowed access on a timed schedule, effectively booking the space and being charged accordingly. Depending on their nature (for people or goods) and location, HODOs could feature shelters, smart columns offering additional services, bike hire facilities and so on. Their location and spacing adjacent to pavements should be carefully considered to maximise accessibility, taking into account the spatial context.

INDUCED DEMAND

Increased use of vehicular infrastructure following its increased provision – in other words the greater the quantity of roads and the larger their capacity, the more traffic will be generated. This widely accepted principle is one of the reasons for avoiding giving over to CAVs road space that may be gained as a consequence of their adoption.

KERB

The edge of the pavement (sidewalk in the US). CAVs will make kerbs the field of revenue and location of spatial battles when street parking is drastically reduced and on-plot parking virtually removed. Especially in city centres, kerbs will have to work hard as flexible spaces used in different ways by different users at different times of the day. The way kerbs are treated is fundamental to the smooth and safe operation of streets, as frequent pull outs for hop-on/drop-off affect other road users including cyclists, other vehicles and pedestrians.

MICROTRANSIT

Transport services normally operated by private companies with vans or small shuttle buses, increasingly offering flexible routes and scheduling enabled by innovative IT applications. Critics of microtransit point out that it undermines public transport by operating on the most popular routes and thus decreasing ridership. Supporters highlight how it identifies under-served areas or routes. With e-hailing services, microtransit provides insights into the benefits and challenges that will be brought by CAVs, particularly as a first/last mile solution.

MOBILITY AS A SERVICE (MAAS)

A form of transport delivered and consumed as a service. It is brokered by mobility providers who identify the best combination of travel options and arrange them for users according to requirements specified via a computer or mobile device. CAVs will appear on MaaS providers' menus and increase the number of travel options. MaaS will become increasingly important in determining the share of people who decide to walk and cycle in the city, influencing the routes they choose as much as factors such as convenience and attractiveness.

MOVEMENT NETWORK

The system of public rights of way catering for movement needs. The introduction of CAVs will reshape the current network but will also change the relative significance of roads and streets for non-movement functions, re-establishing their socio-economic vocations and introducing a further strategic element in cities, particularly an environmental one.

MULTI-LAYERED NETWORK

Transport networks made by combining layers of several networks of transport modes. Layers are separated but they 'touch' at key nodes. In other words, the nodes belong to multiple layers of the network. In the case of CAVs, the nodes are provided by HODOs and NEMOHs, connecting CAV networks to networks for pedestrians, cycling, public transport, and cars. The design and locations of these facilities are crucial to ensuring the CAVs potential is unlocked and that smooth transitions are facilitated by nodes.

- -

MULTIMODAL (TRANSPORTATION)
Transport involving two or more modes within the same trip, such as bicycle plus train plus CAV. The concept of using multiple modes to complete a trip is an extension of the fact that optimal travel arrangements vary across contexts and are provided by different services and technologies, requiring transfers. CAVs will play an essential role in city centres as they will be one of the options to cover first and last miles. Crucial to the success of multimodality and the CAVs strategy is a smooth and comfortable transition between modes, which has strong implications for the way HODOs and NEMOHs are designed and integrated in the urban fabric.

NEW MOBILITY HUBS (NEMOHS)
Transport interchanges featuring CAVs. There will be different types of NEMOHs, which could be categorised as Major, Neighbourhood and Remote depending on their location, level of integration with traditional transport modes, and the functions and services provided (storage, servicing and access). NEMOHs will facilitate the transition from the current car-based system to a multimodal system consisting of walking, cycling, public transport and CAVs.

PARKING STANDARDS
Minimum parking standards are typically required by law for urban development. Mandatory provision of on-site car spaces impacts on achievable densities, and is a key factor in dictating the amount of space available for the actual building, in determining the actual cost of development schemes, and ultimately their viability. As shared CAVs dramatically reduce the demand for car parking, parking standards should be de-bundled

from development requirements. The space gained should be decoupled from vehicular needs for better uses, such as improving amenities and the provision of green infrastructure, SUDs, spill-out space for business and, where appropriate, developing denser communities.

PAVEMENTS (UK) / SIDEWALKS (US)
Pedestrian space. CAVs offer the chance of scaling up the existing infrastructure as street parking space declines and is reallocated in favour of pedestrians and cyclists.

PLATOONING
Use of connectivity technology to enable CAVs to form and maintain close-headway formations. The advantage is to make space savings. Platooning should be carefully managed in urban areas to provide breaks that allow pedestrian crossings and to maintain visual permeability, avoiding the creation of barriers in the middle of the street.

PUBLIC TRANSPORT (UK) OR MASS TRANSIT (US)
The most efficient way to move large numbers of people quickly in a city. Depending on how they are regulated, CAVs will either undermine public transport by competing with it or make it more attractive by complementing it and proving first/last mile solutions that cover areas where a frequent public transport service is too expensive to provide.

RIDE SHARING
E-hailing car service in which the driver is able to make multiple stops and pick up different passengers going in the same direction. The introduction of CAVs and the subsequent creation of HODOs will require a careful

management of the kerb (see above). The convergence of ride sharing with autonomous driving has the potential to shift the predominant automobile ownership model from private ownership to a shared mobility model.

ROAD AND STREET
The Oxford Dictionary defines a road as: 'a wide way leading from one place to another, particularly one with a specially prepared surface which vehicles can use', and as: 'the part of a road intended for vehicles, especially in contrast to a verge or pavement'. A street is defined as: 'a public road in a city, town, or village, typically with houses and buildings on one or both sides.' The road is effectively monofunctional i.e. entirely meant for transport. A street is meant for people not just for movement but also for socio-economic functions. With the right political drive, CAVs will be instrumental in changing roads back into streets in urban areas by reallocating space.

SHARED SPACE
A streetscape design approach eliminating physical separation (horizontal and vertical signage, kerbs, furniture, etc.) between road users as a way to generate positive uncertainty, which acts as a natural traffic calming measure. With CAVs, uncertainty would remain for humans but would not apply to the vehicle, however they would deliver less cluttered environments and minimise barriers, therefore improving accessibility.

SIDEWALKS (US)
see **Pavements**

SMART CITIES
Loose term indicating a range of initiatives adopted by cities using technologies and data to

increase operational efficiency and improve communication and services for their citizens. Many definitions are data-driven and focus on data harvesting and processing to improve decision making and responsiveness. For many, especially manufacturers and solution providers, an essential component of smart cities is a network of sensors permeating the city enabling data collection and monitoring. The CAVs concept can be ambivalent towards smart cities even if they are often associated. On the one hand, CAVs benefit from exchanging information with the surrounding infrastructure, like traffic lights or parking spots, on the other hand they do not necessarily rely on this infrastructure to work. CAVs are designed to work in 'dumb' environments and read external conditions, so CAVs can exist without smart cities. At the same time, to maximise the benefits of CAVs in terms of optimising traffic flows, for instance, CAVs would need to be coordinated by some form of 'central control room', which could be labelled as smart city infrastructure.

SPRAWL
A form of urban expansion based on homogeneous low-density residential neighbourhoods that are heavily reliant on cars and a series of monofunctional hubs where retail or employment are concentrated. Areas of sprawl are often called suburbia. In suburbia public transport is normally not viable given the spread of catchment area. Car space and infrastructure are generous yet don't often resolve congestion problems due to land use zoning. The use of CAVs risks the promotion of further sprawl as it promises a more comfortable or productive commute between urban centres and more distant

and affordable residential areas. A more positive perspective on CAVs sees a solution in densifying suburbia and improving access to public transport by providing sustainable transport options to cover first and last miles from the home to public transportation hubs.

STREET
See **Roads and Streets.**

SUPERBLOCK
Reorganisation of mobility in a section of the city obtained by changing the road network and creating separate routes for different modes of transport. In general, motorised traffic is allowed at urban edges, while inner streets are dedicated to pedestrians and cyclists. Early tests of this approach have been carried out in Barcelona by modifying both the grid network and establishing differentiated routes for each mode of transport. As a consequence, the internal streets of the superblock also change character and promote new relationships with the buildings that face public space. The superblock principles could be applied to a CAVs scenario from the transition phase.

SUSTAINABLE URBAN DRAINAGE SYSTEMS (SUDS)
SUDS mimic nature and typically manage rainfall close to where it falls. SUDS can be designed to convey surface water and to slow run-off before it enters watercourses. They also provide areas to store water in natural contours and can be used to allow water to infiltrate the ground or evaporate from surface water storage or be lost or transpired from vegetation, known as evapotranspiration. CAVs will increase the opportunities to

develop substantial SUDs systems, especially on local streets, thanks to the reallocation of space allowed by a drastic reduction in parking requirements.

TRANSIT / MASS TRANSIT (US)
See public transport (UK).

TRANSITION PHASE
The period in which both CAVs and traditional cars will be used in our cities. The two technologies and the associated ownership and use models could end up competing for space and result in a difficult coexistence. In the attempt to manage this issue, infrastructural rigidities may be created, such as spatial segregations and barriers. As this period may last a generation or more, it could generate a long-lasting legacy for our cities, even many years after traditional cars have disappeared from streets.

TRANSPORTATION-AS-A-SERVICE
See Mobility-as-a-Service.

ZONING
Zoning can be defined as a planning control tool that consists of dividing the land into zones where specific land uses and densities are allowed. The wholesale spatial separation of uses triggers a series of side effects in terms of transport and, in combination with car reliance, becomes a recipe for congestion and environmental issues. In a CAVs world, land use regulations that constrain greenfield development and promote infill may be the primary way to prevent further sprawl and encourage urban revitalisation.

Exploring the implications of new trends in urban mobility, like driverless cars, for urban design and architecture:

Dennis, K., Urry J., *After the Car*, Cambridge, Polity Press, 2009

Herger, M., *The Last Driver's License Holder Has Already Been Born: How Rapid Advances in Automotive Technology Will Disrupt Life As We Know It and Why This is a Good Thing*, Kulmbach, Plassen Verlag, 2017

Kerrigan, D., *Life as a Passenger: How Driverless Cars Will Change the World*, London, London Publishing Partnership, 2017

Levi, T., *The Great Race. The Global Quest for the Car of the Future*, New York City, Simon & Schuster, 2016

Lipson, H., Kurman, M., *Driverless: Intelligent Cars and the Road Ahead*, Cambridge, Massachusetts, MIT Press, 2016

On the relationship between cars and urban planning and design at the different scales:

Alexander, C. et al., *A City is Not a Tree*, Portland, Oregon, Sustasis Press, 2015

Appleyard, D., Lynch, K. and Myer, J., *The View from the Road*, Cambridge, Massachusetts, M.I.T. Press, 1966

Barnett, J., *Urban Design as Public Policy: Practical Methods for Improving Cities*, New York City, Architectural Record Books, 1974

Barnett, J., *Redesigning Cities: Principles, Practice, Implementation*, New York City, Routledge, 2017

Belanger, P., 'Redefining Infrastructure', in Mostafavi, M. et al, *Ecological Urbanism*, Baden, Switzerland, Lars Muller Publishers, 2010

Bell, J., *Carchitecture*, London, Birkhäuser, 2001

Ben-Joseph, E., *Rethinking a Lot: The Design and Culture of Parking*, Cambridge Massachusetts, 2015

Boston Transportation Department, *Go Boston 2030 Vision and Action Plan*, 2017, www.boston.gov, (accessed 15 September 2018)

Campoli, J., *Made for Walking: Density and Neighbourhood Form*, Concord, New Hampshire, Lincoln Institute of Land Policy, 2012

Gehl, J., *Life Between Buildings: Using Public Space*, Washington DC, Island Press, 2011

Greater Vancouver Regional District and the Province of British Columbia, *A Long-range Transportation Plan for Greater Vancouver: Transport 2021 report*, Burnaby (B.C.), Greater Vancouver Regional District, 1993

Hall, P., *Cities of Tomorrow: An Intellectual History of Urban Planning and Design in the Twentieth Century*, Malden, Massachusetts, Wiley-Blackwell, 2002

Hamilton-Baillie, B., *Home Zones – Reconciling People, Places, and Transport: Study Tour of Denmark, Germany, Holland, and Sweden, July to August 2000*, Winston Churchill Memorial Trust Travelling Fellowship, 2000

Hart, J. and Parkhurst, G., 'Driven to Excess: A Study of Motor Vehicle Impacts on Three Streets' in Bristol UK, *World Transport Policy and Practice*, 17 (2). pp 12–30, 2011

Kayden, J.S. and The New York City Department of City Planning, *Privately Owned Public Space: The New York City Experience*, New York, John Wiley & Sons, 2000

Krieger, A., 'Where and How Does Urban Design Happen?', *Harvard Design Magazine No. 24: The Origins and Evolution of Urban Design, 1956–2006*, Spring/Summer 2006

Lindquist, E., 'Moving Toward Sustainability: Transforming a Comprehensive Land Use and Transportation Plan', *Land Use and Transportation Planning and Programming Applications*, Volume 1617, 1998

Jacobs, A., *Great Streets*, Cambridge, Massachusetts, The MIT Press, 1993

Jacobs, J., *The Death and Life of Great American Cities*, New York, Vintage Books, 1992

Lynch, K., *The Image of the City*, Cambridge, Massachusetts, The MIT Press,1960

Mumford, L., *The City in History*, New York, Harcourt, Brace & World, 1961

NACTO, *Urban Street Design Guide*, Washington, D.C., Island Press, 2012

Pavia, R., *Il Passo della Città. Temi per la Metropoli Future*, Rome, Donzelli Editore, 2015

Penalosa, E., 'Politics, Power, Cities', in Burdett, R. et al, *The Endless City*, London, Phaidon Press, 2008

Portland, OR (The City of), *Draft Connected and Autonomous Vehicles Policy*, 2017, www.portlandoregon.gov, (accessed 8 September 2018)

San Antonio, TX (The City of), *SA Tomorrow Multimodal Transportation Plan*, 2016, www.satransportationplan.com, (accessed 8 September 2018)

Seattle Department of Transportation, '*Appendix C: Preliminary Automated Mobility Policy Framework*' in *New Mobility Playbook*, 2017, www.seattle.gov, (accessed 8 September 2018)

San Jose, CA (The City of), *Smart City Vision*, www.sanjoseca.gov, (accessed 8 September 2018)

Sennett, R., 'Civility', in *Urban Age*, Bulletin 1, Summer 2005

Speck, J., *Walkable City: How Downtown Can Save America, One Step at a Time*, New York City, North Point Press, 2013

UNFPA, 'Peering into the Dawn of an Urban Millennium', in *State of the World Population 2007*, pp. 1–4, New York, UNFPA Editions, 2007

Urry, J., *Sociology Beyond Societies: Mobilities for the Twenty-first Century*, London and New York, Routledge, 2000

Urry, J., Leach, J., Dunn, N., Coulton, C. and the Liveable Cities Team, *The Little Book of Car Free Cities*, Lancaster, Imagination Lancaster, 2017

Vancouver (The City of), *The City of Vancouver Plan: Transportation 1997*, Vancouver (B.C.) Engineering Services, 1997

Consultancy and public agency short papers and web articles:

Accenture, 'Autonomous Vehicles: Plotting a Route to the Driverless Future', www.accenture.com/gb-en/insight-autonomous-vehicles, (accessed 10 September 2018)

Audi, 'Audi Urban Future Initiative', www.audi-urban-future-initiative.com/blog/, (accessed 10 September 2018)

Bouton, S., Hannon, E., Knupfer, S., and Ramkumar, S., 'The Future(s) of Mobility: How Cities can Benefit', McKinsey, June 2017, www.mckinsey.com, (accessed 5 September 2018)

Bouton, S., Knupfer, S. M., Mihov, I., and Swartz S., 'Urban Mobility at a Tipping Point', McKinsey, September 2015, www.mckinsey.com, (accessed 5 September 2018)

Budds, D., 'What Happens When Lyft Redesigns A Street', *Fast Company*, 9 September 2017, www.fastcompany.com, (accessed 5 September 2018)

Buro Happold, 'Connected and Autonomous Vehicles', www.burohappold.com/the-lab, (accessed 10 September 2018)

Cohen, A., 'The Game-Changer for Future Cities: Driverless Cars', Gensler, Dialogue No.30: The Livability Issue, www.gensler.com/research-insight, (accessed 10 September 2018)

Crute, J., Riggs, W., Chapin, T., Stevens, L., *Planning for Autonomous Mobility*, PAS Report 592, American Planning Association, www.planning.org/publications, (accessed 15 November 2018)

ENO (The Eno Center for Transportation), 'Adopting and Adapting: States and Automated Vehicles', 1 June 2017, www.enotrans.org, (accessed 12 March 2018)

Hobbs, A., Harriss, L., Parliamentary Office of Science and Technology (POST), 'Peak Car Use in Britain', Commons Transport Select Committee, November 2013, www.parliament.uk, (accessed 10 August 2018)

Howarth, D., 'Foster + Partners and Nissan Unveil Vision for Self-charging Driverless Cars that can Power the Home', *Dezeen*, 1 March 2016, www.dezeen.com, (accessed 5 September 2018)

KPMG, 'Parking Demand in the Autonomous Vehicle Era', 17 July 2017, www.kpmg.com, (accessed 10 November 2018)

Litman, T., *Autonomous Vehicle Implementation Predictions: Implications for Transport Planning,* Victoria Transport Policy Institute, 26 November 2018, www.vtpi.org/avip.pdf, (accessed 10 August 2018)

NACTO, *Blueprint for Autonomous Urbanism*, www.nacto.org/publication/, (accessed 15 September 2018)

National Academies of Sciences, Engineering and Medicine, Transportation Research Board, National Cooperative Highway Research Program, Gettman, D., Lott J., S., Goodwin, G., Harrington, T., *Impacts of Laws and Regulations on CV and AV Technology: Introduction in Transit Operations*, NCHRP, 2017, www.nap.edu/catalog/24922/, (accessed 15 September 2018)

NHTSA, 'Automated Vehicles for Safety', www.nhtsa.gov/technology-innovation/, (accessed 5 September 2018)

Parsons Brinckerhoff and Center for Automotive Research, *Use of Data from Connected and Automated Vehicles for Travel Demand Modeling*, Michigan Department of Transportation (MDOT), October 29 2015, https://www.michigan.gov/documents/, (accessed 20 August 2018)

RAC Foundation, 'Council Parking Revenue in England 2016–17', 27 Nov 2017, www.racfoundation.org/research/, (accessed 20 August 2018)

RAC Foundation, 'Spaced Out: Perspectives on Parking Policy', 17 Jul 2012, https://www.racfoundation.org/research/, (accessed 20 August 2018)

Reese, H., US DOT unveils 'world's first autonomous vehicle policy,' ushering in age of driverless cars', *Tech Republic*, September 20, 2016, www.techrepublic.com, (accessed 10 August 2018)

Regional Plan Association, *New Mobility: Autonomous Vehicles and the Region*, 9 October 2018, www.rpa.org/publication/, (accessed 20 August 2018)

UK Government Guidance, 'The Key Principles of Vehicle Cyber Security for Connected and Automated Vehicles', 6 August 2017, www.gov.uk/government/publications/, (accessed 21 March 2019)

WSP, Parsons Brinckerhoff, Farrells, *Making Better Places: Autonomous Vehicles and Future Opportunities*, www.wsp-pb.com, (accessed 5 September 2018)

Newspaper and blog articles:

Baldwin, R., 'It Takes a Smart City to Make Cars Truly Autonomous', 14 June 2017, www.engadget.com, (accessed 8 September 2018)

Bilger, B., 'Auto Correct: Has the Self-driving Car at Last Arrived?', *The New Yorker*, 25 November 2013, www.newyorker.com/magazine, (accessed 6 September 2018)

Brake, A. G., 'MIT Researchers Plan 'Death of the Traffic Light' with Smart Intersections', *Dezeen*, 21 March 2016, www.dezeen.com, (accessed 8 September 2018)

Davies, A., 'The Very Human Problem Blocking the Path to Self-driving Sars', *Wired*, 1 January 2017, www.wired.com, (accessed 6 September 2018)

Essbai, S., 'Autonomous Driving: Are Cities Ready?', www.theglobalgrid.org, 10 January 2018, www.theglobalgrid.org, (accessed 6 September 2018)

Garret, O., '10 Million Self-Driving Cars Will Hit The Road By 2020 – Here's How To Profit', Forbes, 3 March 2017, www.forbes.com/sites/oliviergarret/, (accessed 6 September 2018)

Gartland, M., Jaeger, M., 'GM to Start Testing Driverless Cars in NYC', *New York Post*, 17 October 2017, (accessed 8 September 2018)

Greenwood, G., 'Driverless Cars Pose Threat to Growth of Cycling in Cities', *Financial Times*, 6 November 2017, www.ft.com, (accessed 8 September 2018)

Geeting, J., 'It's an Automatic: The Road to a Future of Driverless Cars, Dense Streets and Supreme Mobility', Nextcity.org, 10 February 2014, www.nextcity.org/features/, (accessed 5 September 2018)

Godsmark, P., 'The Definitive Guide to the Levels of Automation for Driverless Cars', Driverless.wonderhowto.com, 4 September 2017, www.driverless.wonderhowto.com, (accessed 6 September 2018)

Harris, J., 'Owning a Car Will Soon be a Thing of the Past', *The Guardian*, 23 October 2017, www.theguardian.com, (accessed 6 September 2018)

Hodgkinson, T., 'We Live in an Age of Disruption. I'd Rather be Creative', *The Guardian*, 29 September 2015, www.theguardian.com, (accessed 6 September 2018)

Laris, M., Halsey, A., 'Will Driverless Cars Really Save Millions of Lives?' *Lack of Data Makes it Hard to Know, The Washington Post*, 18 October 2016, www.washingtonpost.com, (accessed 8 September 2018)

Mehra, V., 'The Evolution of Parking in a World of Driverless Cars', Parquex.com, 26 April 2017, www.parqex.com, (accessed 6 September 2018)

Moss, S., 'End of the Car Age: How Cities are Outgrowing the Automobile', *The Guardian*, 28 April 2015, www.theguardian.com/cities, (accessed 6 September 2018)

Newcomb, D., 'How Driverless Cars Spell the End of Parking as We Know It', *PCmag*, 12 August 2016, www.uk.pcmag.com, (accessed 6 September 2018)

Ng, A., Lin, Y., 'Self-driving Cars Won't Work Until we Change our Roads and Attitudes', *Wired*, 15 March 2016, www.wired.com, (accessed 8 September 2018)

Rapier, G., 'Bank of America: We've Reached "Peak Car"', *Business Insider*, 15 June 2017, www.uk.businessinsider.com, (accessed 6 September 2018)

Renn, A., 'Have we Really Reached "Peak Car"?' *The Guardian*, 30 April 2015, www.theguardian.com/cities/, (accessed 6 September 2018)

Reese, H., 'Updated: Autonomous Driving Levels 0 to 5: Understanding the Differences', Techrepublic.com, 20 January 2016, www.techrepublic.com, (accessed 6 September 2018)

RethinkX, 'New Report: Due to Major Transportation Disruption, 95% of U.S. Car Miles will be Travelled in Self-driving, Electric, Shared vehicles by 2030', www.rethinkx.com, (accessed 6 September 2018)

Revesz, R., 'Jeremy Clarkson Warns that Driverless Cars are Dangerous', *The Independent*, 19 November 2017, www.independent.co.uk, (accessed 5 September 2018)

Schneider, B., 'Do Driverless Cars Need Their Own Roads Around Manhattan?', 26 July 2017, Citylab.com, www.citylab.com/transportation/, (accessed 5 September 2018)

Schwanen, T., 'Peak Car? Driverless Technology May Actually Accelerate Car Ownership', *The Guardian*, 18 July 2017, www.theguardian.com/sustainable-business/, (accessed 6 September 2018)

Skorup, B., 'Driverless Cars Need Just One Thing: Futuristic Roads', *Wired*, 10 October 2016, www.wired.com, (accessed 5 September 2018)

Sohrweide, T., 'Driverless Vehicles Set to Change the Way We Design Our Roadways?', Short Elliott Hendrickson Inc , 25 July 2018, http://www.sehinc.com/news/future-what-do-driverless-cars-mean-road-design, (accessed 5 September 2018)

Soudi, A., 'Driverless Cars Might Follow the Rules of the Road, But What About the Language of Driving?', *The Conversation*, 8 January 2018, www.theconversation.com, (accessed 8 September 2018)

Speck, J., 'Ten Rules for Cities About Automated Vehicles', *Medium*, 16 October 2017, www.medium.com, (accessed 6 September 2018)

The Economist, 'Driverless Cars,' 3–9 March 2018, p 53

Van Mead, N., 'America's Road Trip: Will the US Ever Kick the Car Habit?', *The Guardian*, 2 November 2016, www.theguardian.com/cities, (accessed 6 September 2018)

Walker, T., '"It's the Worst Place to Park in the World" – Why Britain is at War Over Parking", *The Guardian*, 29 May 2017, www.theguardian.com/world (accessed 6 September 2018)

Wolmar, C., 'The Myth of the Driverless Cars Revolution', 11 April 2016, www.christianwolmar.co.uk, (accessed 5 September 2018)

Wolmar, C., 'Transport's Favourite Myth: Why we Will Never Own Driverless Cars', *New Statesmen*, 10 April 2016, www.newstatesman.com/culture, (accessed 8 September 2018)

Young, A., 'Self-driving Cars vs. American Roads: Will Infrastructure Speed Bumps Slow Down the Future of Transportation?', *Salon,* 20 April 2017, www.salon.com, (accessed 8 September 2018)

Regarding London specifically:

Buchanan, C., *Traffic in Towns: The Specially Shortened Edition of the Buchanan Report*, London, Penguin Books, 1963

Inrix, 'New Inrix Study Reveals Car Traffic in London is Down but Congestion is Up', London, 17 May 2016, www.inrix.com/press-releases/london-traffic/, (accessed 10 September 2018)

Laker, L., 'Street Wars 2035: Can Cyclists and Driverless Cars Ever Co-exist?', *The Guardian*,14 June 2017, www.theguardian.com/cities, (accessed 5 September 2018)

Morley, K., 'Record Decline in Teenagers Learning to Drive, Figures Show', *The Telegraph*, 10 July 2017, (accessed 5 September 2018)

Roads Task Force, 'To What Extent is Congestion and Unreliability on the Road Network Caused by Factors that can be Influenced by TfL's Road Network Management?', TfL, Technical Note 11, 2013, www.tfl.gov.uk/technical-note-11, (accessed 10 September 2018)

Roumpani, F., Hudson, P., 'The Evolution of London: The City's Near-2,000 Year History Mapped', *The Guardian*, 15 May 2014, www.theguardian.com/cities, (accessed 8 September 2018)

Transport Committee, *London Stalling. Reducing Traffic Congestion in London,* London Assembly, January 2017, www.london.gov.uk, (accessed 8 September 2018)

Transport for London, *London's Transport – A History,* www.tfl.gov.uk, (accessed 10 September 2018)

Transport for London, *Strategic Cycling Analysis. Identifying Future Cycling Demand in London*, June 2017, www.tfl.gov.uk, (accessed 10 September 2018)

Transport for London, *Travel in London – Report 9*, 2016, www.tfl.gov.uk, (accessed 10 September 2018)

Regarding Los Angeles specifically:

Banham, R., *Los Angeles: The Architecture of Four Ecologies*, London, Penguin Books, 1973

Bradbury, R., 'The Aesthetic of Lostness', in Bradbury, R., *Yestermorrow: Obvious Answers to Impossible Futures*, Santa Barbara, CA, Joshua Odell Editions, 1988

Bradbury, R., Weller, S., *Ray Bradbury: The Last Interview and Other Conversations*, Melville House, Brooklyn, NY, 2014

Davis, M., *City of Quartz: Excavating the Future in Los Angeles*, London, Verso, 2006

Department of City Planning Los Angeles, *The Concept for the Los Angeles General Plan*, 1970

Discover Los Angeles, *Facts About Los Angeles*, 15 December 2017, www.discoverlosangeles.com/press-releases, (accessed 20 August 2018)

Fain, W., *If Cars Could Talk*, Balcony Press, Glendale, California, 2012

Fulton, W., The Reluctant Metropolis: The Politics of Urban Growth in Los Angeles, Baltimore, Maryland, Johns Hopkins University Press, 2001

Grad, S., 'Los Angeles Hits a Milestone: 4 Million People and Counting', *Los Angeles Times*, 2 May 2017, www.latimes.com, (accessed 20 August 2018)

Los Angeles Economic Development Corporation, Los Angeles County Profile, www.laedc.org/reports/, (accessed 20 August 2018)

LA Almanac, 'First Automobile in Southern California', www.laalmanac.com/transport/, (accessed 20 August 2018)

Manville, M., Soup, D., 'People, Parking and Cities', 234 / *Journal of Urban Planning and Development*, December 2005, www.researchgate.net, (accessed 20 August 2018)

METRO, *Measure M*, Los Angeles, www.theplan.metro.net, (accessed 20 August 2018)

Southern California Association of Governments, *The 2016–2040 Regional Transportation Plan/ Sustainable Communities Strategy*, April 2016, www.scagrtpscs.net, (accessed 20 August 2018)

Urbanmobilityla, *Urban Mobility in a Digital Age, August 2016,* Goldhirsh Foundation in partnership with the Mayor's Fund of Los Angeles, www.urbanmobilityla.com, (accessed 20 August 2018)

REFERENCES

PREFACE

1   For current market predictions see the USA National Highway Traffic Safety Administration (NHTSA), www.nhtsa.gov

CHAPTER 1

1   Clarke, A.C., 2000, *Profiles of the Future* (2nd rev. ed.) Indigo, London, p33.

2   Preston, Schiller and Kenworthy, 2017, *An Introduction to Sustainable Transportation*.

3   https://www.oldbaileyonline.org/static/Transport.jsp

4   Diagram based on population Census data (https://data.london.gov.uk) and the London Evolution Animation (www.theguardian.com) produced by CASA (www.ucl.ac.uk/bartlett/casa).

5   Preston, Schiller and Kenworthy, 2017, An Introduction to Sustainable Transportation, p 41.

6   www.britannica.com/technology/automobile

7   According to the LA Almanac: www.laalmanac.com

8   Dennis and Urry, 2009, *After the Car*.

9   Norton, P., 2008, *Fighting Traffic: The Dawn of the Motor Age in the American City*, The MIT Press, Cambridge, Massachusetts, London, England.

10 Shoup, D., *The High Cost of Free Parking*.

11 Even today and even in countries with higher gas taxes like Germany, motorists still don't fully pay for the externalities they create.

12 The love for the automobile and its role in consumerism spread quickly from America across the globe. In *For Love of the Automobile: Looking Back Into the History of Our Desires* Wolfgang Sachs perfectly captured the nature of phenomenon in Germany.

13 Mr Sloan made the year of production of the car recognisable, as for a dress or another object of fashion. By doing so he deployed a planned obsolescence strategy of the car.

14 Referring to the car as a product is a huge simplification. In the words of Dennis and Urry in After the Car, the car is 'a node within a broader system' whose impacts 'stem from its entire life cycle and related infrastructure systems, including extraction of raw materials, vehicle production, operation and maintenance, as well as maintenance of the road infrastructure, hospital costs, emotional costs of the many deaths and injuries'.

15 See Kerrigan, 2017, *Life As A Passenger: How Driverless Cars Will Change The World*, p57: 'The percentage of Americans holding driver's licenses has fallen sharply over the past several decades, especially among the young. In 1983, more than 91 percent of 20-to-24-year-olds held a license. By 2014, that number had dropped to approximately 77 percent and shows little sign of recovering. [...] Younger generations, the ones who grew up with game consoles and smartphones, are not so in love with cars. This group-members of the "Millennial" generation-are not rushing to get driver's licenses the way baby boomers did. They live perpetually connected lives, and while they may have the same desire for mobility on demand, some see the act of driving as a distraction from texting, not the other way around.'

16 KPMG Global Automotive Executive Survey 2017.

17 See for instance 'Five Trends Transforming the Automotive Industry', PricewaterhouseCoopers, 2018.

18 The UN's *The Global Mobility Report* (2017) is the first ever attempt to examine the performance of the transport sector globally.

19 The numbers are also daunting when looking at the environmental consequences of this trend, as the 2017 *The Global Mobility Report* explains.

20 Moses R., 1962, 'Are Cities Dead?', *The Atlantic*, January 1962 Issue. See www.theatlantic.com

21 The assumptions are setting the average length of a car at 4.5m and excluding vans and trucks used to transport goods.

22 See 'Today's Cars are Parked for 95 Percent of the Time', 13 March 2016, www.fortune.com

23 Peter Hall in *Cities of Tomorrow* (pp 347–48) points out that it's impossible to say which came first, the 'suburban chicken or the automotive egg' as suburban sprawl predated mass car ownership both in London and Los Angeles. However, the automobile allowed the suburbs to sprawl more freely than mass transit could ever have done.

24 Duany, A., 2001, *Suburban Nation: The Rise Of Sprawl And The Decline Of The American Dream*.

25 Jacobs, J., 1961, *The Death and Life of Great American Cities*.

26 Hall, P., *Cities of Tomorrow*, p 334.

27 *Traffic in Towns: The Specially Shortened Edition of the Buchanan Report*, 1964.

28 *In The City Assembled*, Kostof observes that 'the logic of circulation has not always guided city-making. Residential streets of status could elect a grand scale incommensurate with their modest traffic flow. And paths of main commercial activity were often near-impassable bottlenecks, willingly tolerated on the premise that crowds in tight quarters were the key to contagious shopping. Street-widening in the heavily congested city centre of the early automobile years was not always welcome by the store owners who were considered its primary beneficiaries. [...] The consensus of the time was that main traffic streets offered the best opportunities for shops provided they were not wider than about 50 to 70 feet (15–20 m)'.

29 See Kostof, p 234.

30 Buchanan, 1964, *Traffic in Towns*.

31 900 shopping malls were built in around three decades in the UK (source www.theguardian.com), while in the US 1,500 malls were built between 1956 and 2005, 'and their rate of growth often outpaced that of the population' (source: www.tlme.com).

32 See for instance the work of Appleyard, Lynch and Mycr in the 1965 study *View from the Road*, or 1972's *Learning from Las Vegas* by Venturi, Scott Brown and Izenour.

33 Source: the Webster dictionary, www.merriam-webster.com

34 Rogers, R., 1997, *Cities for a Small Planet*.

35 Urry, J., 2000, *Sociology beyond societies*.

36 Source: www.nytimes.com, Towns against Traffic, 1972 pp.14-1519 March 2018.

37 https://newrepublic.com/article/150689/modern-automobile-must-die

38 Caiazzo, F., Ashok, A., Waitz, I., Yim, S., Barrett, S., 2013, "Air pollution and early deaths in the United States. Part I: Quantifying the impact of major sectors in 2005", MIT

39 From the documentary *Social Life of Small Urban Spaces*, 1979

40 https://www.nytimes.com/2018/03/19/technology/uber-driverless-fatality.html

41 Kerrigan, D., 2017, *Life as a Passenger: How Driverless Cars Will Change The World* reports that in 1867 in New York horses were killing an average of four pedestrians a week, p 36.

42 Good initial points of reference are Kerrigan, D., 2017, *Life as a Passenger* and Lipson, H., Kurman, M., 2016, *Driverless: Intelligent Cars and the Road Ahead*, MIT press.

43 Lipson, H., Kurman, M., 2016, *Driverless*, MIT press, p 107.

44 Ibid, p149.

45 Lipson, H., Kurman M., 2016, *Driverless* and Kerrigan, D., 2017, *Life As A Passenger* give good accounts of the technological advances that are making CAVs possible.

46   In this sense by Smart City we mean 'one that makes optimal use of all the interconnected information available today to better understand and control its operations and optimize the use of limited resources' (source Cosgrove, M. et al, 2011, *Introducing the IBM City Operations and Management Solutions*, IBM Smart Cities series). We have opted for a data-driven definition of the Smart City, rather than a broader or citizen-focused definition, because, rightly or wrongly, this tends to be the dominant one in the debate. See also the glossary.

47   Greenfield, A., 2013, *Against the Smart City*, Do Projects.

CHAPTER 2

1   PAS Report 592, *Planning for Autonomous Mobility*, p 16.

2   Ibid

3   See www.waymo.com/tech/

4   See glossary for a definition of ATMS.

5   See 'Navigant Research Leaderboard: Automated Driving Vehicles', www.navigantresearch.com

6   See ENOTRANS 'Adopting and Adapting: States and Automated Vehicles' for the current state of policies related to automated driving in the US, www.enotrans.org

7   Level 4 relates to full automation in controlled areas, in which there will be no need for a driver, and the vehicle will be able to handle emergencies and, in case of problems, safely pull over and stop.

8   'The IHS predicts 21 million fully or semi-autonomous sold globally in 2035 and a total of 76 million sold between now and then.' Source: ihsmarkit.com

9   Mosquet et al, 2015; McKinsey & Company, 2016; Walker Consultants, 2017.

10   Source: www.un.org

11   Source: www.businessinsider.de

12   See 'Urbanites Flee China's Smog for Blue Skies', 22 November 2013, www.nytimes.com

13   See regarding this debate Richard Florida's work in *The Rise of the Creative Class: And How It's Transforming Work, Leisure, Community, and Everyday Life*, and more recently *The New Urban Crisis: How Our Cities Are Increasing Inequality, Deepening Segregation, and Failing the Middle Class—and What We Can Do About It*.

14   Fagnant, Kockelman and Bansal, 2015.

15   According to NHTSA, the safety benefits of automated vehicles will be paramount. Automated vehicles will have the potential to remove human error from the crash equation, which will help protect drivers and passengers, as well as cyclists and pedestrians. According to NHTSA more than 35,092 people died in motor vehicle-related crashes in the US in 2015.

16   For further reading regarding market predictions see 'The Self-driving Car Timeline: Predictions from the Top 11 Global Automakers', www.techemergence.com

17   The current average age of cars is 10.7 years in the EU (source: European Automobile Manufacturers Association, 2017) and 11.6 years in the US (source: IHS Markit, 2016).

18   This assumption implies the end of a trend inaugurated 40 years ago when the average car width increased by 16 per cent. (Source: www.dailymail.co.uk). Change is likely to be imposed only by regulation and will have to satisfy safety requirements..

19   For more details on forecasts for parking demand in the autonomous vehicle era, see Steer's work in collaboration with KPMG at www.kpmg.com

20   Vehicle-to-vehicle (V2V) communication's ability to wirelessly exchange information about the speed and position of surrounding vehicles.

21   It is worth noting that cars today are parked for on average 95 per cent of the time (source: www.reinventingparking.org).

22   The vast majority of US commuters (76.3 per cent) continue to drive to work alone in their cars. Regarding latest commuting trends see www.usnews.com

CHAPTER 3

1   'An Integrated Perspective on the Future of Mobility', www.mckinsey.com and 'Shape Shifting Cities', www.kpmg.com. For a blueprint on autonomous urbanism see www.nacto.org

2   The drop-off section of the kerb, or bays, where people and/or goods access CAVs

3   See for example 'Cycling to Work has Substantial Health Benefits, Study Finds', 20 April 2017, www.bmj.com

4   For more details regarding platooning see also at the white paper by ATA Technology and Maintenance Council: *Automated Driving and Platooning Issues and Opportunities*

5   According to statistics from Strava users, the average speed of a cyclist in London is approximately 14 mph (source: www.bikeradar.com). We believe bringing the speed of vehicles and cyclists closer by reducing the speed of cars is the first step towards safety, more convivial spaces, a smoother operation of the network and ensuring cycling stays at the top of the transport mode hierarchy – coming only after walking and before public transport, CAVs and cars.

6   This applies to cities that did not undergo Baroque, Haussmannian or fascist/communist reconfigurations. In this sense London is probably one of the exceptions among large European cities and a 'worst case' scenario, which makes it interesting to study and compare with its opposite, Downtown Los Angeles. This widens the study's span of relevance, as most central urban areas fall between the two extremes.

7   For more details regarding this strategy see 'Superblocks Rescue Barcelona' at www.theguardian.com/cities

8   This could be an evolution of the trend inaugurated recently in central London by delivery companies like DPD. See 'DPD Opens First Electric Parcel Depot in Central London' at www.greenfleet.net

9   At least on the streets, as it is possible to anticipate the use of drones in urban areas for emergency services and, over time, potentially for logistics and eventually for people.

10   Researchers from the University of Toronto Faculty of Applied Science & Engineering find that optimising for autonomous vehicles could increase the capacity of a parking lot by 62 per cent. See 'How Self-driving Cars Could Shrink Parking Lots', www.sciencedaily.com

11   On this topic see also the seminal work by Kayden, J.S., in *Privately Owned Public Space: The New York City Experience*, 2000.

12   The current impact of logistics on city traffic is exemplified by the case of the UK urban street where one in five trips is made by a van (source: www.gov.uk).

13   See for instance recent extreme climate events in the UK such as the December 2016 flooding, 'Last Winter's Floods Most Extreme on Record', and Summer 2017 heatwaves, 'UK Heatwave to Set 40-Year Temperature Record', both www.theguardian.com

14   See in this regard the US flooding emergency in Houston, 'Exploring Why Hurricane Harvey Caused Houston's Worst Flooding', www.npr.org. See also 'LA's October Heatwave was Shattering Temperature Records', www.dailynews.com and 'Local Wildfires', www.latimes.com

CHAPTER 4

1   See London data on mobility at https://data.london.gov.uk/dataset/licensed-vehicles-type-0

2   See Southern California data on mobility at http://scagrtpscs.net/SiteAssets/ExecutiveSummary/index.html

3   Sources: Greater London Authority and Office for National Statistics.

4   Sources: US Census; City of Los Angeles Mobility Plan 2035; LADOT – The City of Los Angeles Transportation Profile; and 'Los Angeles Mobility Analysis'

5   Sources: Greater London Authority and Office for National Statistics.

6   Green Belt policies aim at stopping urban sprawl by creating protected areas surrounding urban centres to be kept undeveloped or as agricultural land. Famously, the UK introduced a policy in 1955 that applies to a number of its towns and cities.

7   Especially in East and South London. See for instance 'An Urban Renaissance Achieved?' at rhttps://citygeographics. org and London First & Savills, 2015, *Redefining Density: Making the Best Use of London's Land to Build More and Better Homes*. A discussion on the construction rate and the prolonged housing crisis is beyond the scope of this work, however they are crucial topics that also determine the form and type of the densification process as well as its long-term implications.

8   An article in *The Guardian* in 2016 reported a study that attempted to quantify the savings in house prices per minute per train journey from central London. It concluded with a figure of £3,000 for every minute of commute.

9   Arup, 2017, *London's Strategic Infrastructure Requirements: An Evidence Base for the London Plan*, GLA.

10 The management of the road network in London falls under the responsibility of several different organisations: Transport for London (TfL) deals with the strategic network (also known as TLRN or 'red routes') while the 32 London boroughs and the City of London manage the remaining roads within their administrative boundaries. In addition, Highways England is responsible for the motorway network (M25, M1, M4 and M11) in outer London.

11 'TfL has calculated for each of the years for which speed and flow data are available an index of estimated road network capacity based on the position in 1996 [...] It is seen that capacity reduced most in central London – down by about 30 per cent. The loss of capacity in this area as in inner and outer London has been fairly steady, other than around the introduction of congestion charging in 2003. In inner London the loss was about 15 per cent, with an approximate 5 per cent loss in outer London. This reduced capacity has impacted on levels of service on the road network, reducing average traffic speeds and increasing congestion.' Transport for London, Roads Task Force: Technical Note 10, 2013.

12 According to INRIX's annual report on London's traffic (www.inrix.com), journey times in Central London are increasing by 12 per cent annually. The report also highlights the main drivers of this deterioration – planned roadworks, unplanned incidents and a booming e-commerce market affecting the amount of delivery vans on the street. The report also points out that private hire vehicles (including taxis and Uber) are not key contributors to increased traffic congestion.

13 Daily trips in London have increased 130 per cent since the year 2000 (TfL, June 2017, Strategic *Cycling Analysis: Identifying Future Cycling Demand in London*.

14 A 20 per cent drop in under-25s learning to drive has been recorded in the UK (source: www.telegraph.co.uk).

15 In 2016, 43 per cent of households did not have a car. However, in inner London the figure is substantially lower (source: TfL, 2016, *Travel in London: Report 9*).

16 'It was an ambitious network with multiple concentric 'A', 'B' and 'C' ring-roads running through the built-up area, and an outer 'D' orbital in the green belt. The last of these roads was completed in 1986 as the M25 motorway. The others died a slow political death. (Hebbert, M., 2001, Wiley, London, p 75).

17 For a short history of Los Angeles see www.britannica.com

18 See US Census Bureau data for population growth, www.census.gov

19 According to the LA Almanac, 'First Automobile in Southern California', www.laalmanac.com

20 'Los Angeles Hits a Milestone: 4 Million People and Counting', www.latimes.com

21 'Facts About LA', www.discoverlosangeles.com

22 'Los Angeles–Long Beach–Anaheim CA Metro Area' https://datausa.io

23 According to Bloomberg, https://www.bloomberg.com/graphics/infographics/densest-cities-in-2025.html

24 In this regard see SCAG's 2016–2040 Regional Transportation Plan at www.scagrtpscs.net

25 See Lewis Mumford, *The City in History*, 1961, p 510.

26 Manville, M., and Shoup, D., 'Parking, People, and Cities', 234 / *Journal of Urban Planning and Development*, December 2005.

27 Vision Zero is a strategy to eliminate all traffic fatalities and severe injuries while increasing safe, healthy, equitable mobility for all. It is an initiative involving a network of cities in the UK, and takes inspiration from a similar 1990s programme from Sweden subsequently implemented in Europe. Today, this approach is gaining momentum in major American cities.

28 See METRO's 'Plan for Measure M' at: http://theplan.metro.net

29 'LA's Big Plan to Change the Way We Move', www.curbed.com. See also www.urbanmobilityla.com/strategy/

30 See work by NACTO 'Automated Vehicles and the Future of City Streets', www.nacto.org

31 The 2016–2040 Regional Transportation Plan Executive Summary, www.scagrtpscs.net

CHAPTER 5

1   For more information on the origins of Shoreditch see South Shoreditch Conservation Area Appraisal at www.hackney.gov.uk

2   The Congestion Charge is a weekday, daytime daily toll charge for vehicles within central areas.

3   In the UK the average age of a car at scrappage in 2015 was 13.9 years (source: 2017 Automotive Sustainability Report, The Society of Motor Manufacturers and Traders).

4   A solution allowing customers to book various forms of transportation via a mobile device.

5   As discussed in Chapter 4.

6   According to data from the Strava Year in Sport UK 2017, the average ride to work 22.1 kmh (13.7 mph).

7   Inductive or wireless charging is the use of electromagnetic field to transfer energy between charging stations and vehicles without the need for connecting cables. It applies to parked vehicles as well as during movement.

8   See also Hackney Council's Low Emission Neighbourhood Plans at www.zeroemissionsnetwork.com

9   A 3-metre lane is considered substandard in London, however the reduction of traffic and its nature (restricted) and dedicated space for cyclists may justify a potential deviation from current standards.

10 An additional 0.2 m may be required for emergency vehicles bringing the total width usable by an emergency vehicle to 3.7 m. However, this space can be provided on the pavement along the road space.

11 For details on the calculations refer to case study 'What If Cars Were Banned From The Streets?' in this publication.

REFERENCES

12 Assuming each tree absorbs 5.9 kg (or 13 pounds) of CO_2 per year, and that they are young trees. Information used for the calculation sourced from www.urbanforestrynetwork.org

13 An important limitation to our exercise relates to the lack of a reliable travel demand forecast models for CAVs (none exist as CAVs are still to be introduced) and the specificity of the London context in terms of modal share and user preferences. Therefore a number of assumptions have been made, including: no behavioural changes take place following the introduction of CAVs; distribution of trips within the hours is even (the granularity of our peak time forecast data is hourly); each trip rate is limited to the same land use in town- and neighbourhood-centre locations; modal split is based on Census 2011 Travel to Work data; waste generation used to calculate servicing needs is based on professional judgement and latest policy; CAVs and service vehicles continue to require bays of 6 x 2 metres (19.6 x 6.5ft); CAVs are not private but operate as a MaaS; and the average time CAVs occupy the drop-off point is 3 minutes. The latter seems reasonable considering current trends in taxi and rideshare services, which have a tolerance of 3 to 5 minutes before charging for the waiting time. Longer time would imply higher charges therefore discouraging longer use of the HODOs.

14 In recent years in Central London virtually all new developments already have zero parking provision (except a quota of disabled car parking).

CHAPTER 6

1 See regarding this pivot vision plan for Los Angeles a recent article from *Planetizen*, www.planetizen.com/node/23535

2 'Philadelphia Grid Marks Birth of America's Urban Dream', www.theguardian.com/cities/

3 DCBID Annual Report 2016 www.downtownla.com

4 See regarding this topic 'Just Because You Can't Find a Place to Park', January 14, 2015, www.citylab.com

5 See complete report at: http://www.transportationlca.org/losangelesparking/

6 Manville, M., and Shoup, D., 'Parking, People, and Cities', 234 / *Journal of Urban Planning and Development*, December 2005.

7 See 'Automated Vehicles for Safety' from NHTSA for latest market predictions, www.nhtsa.gov

8 https://www.ieee.org/about/news/2012/5september_2_2012.html

9 'Ten Ways Autonomous Driving Could Redefine the Automotive World', www.mckinsey.com

10 On this topic see also Joshua Hart's *Driven to Excess: A Study of MotorVehicle Impacts on Three Streets in Bristol UK.*

11 For more details on SDG study on CAVs future parking demand see 'Parking Demand in the Autonomous Vehicle Era', https://home.kpmg.com

12 See on this topic the seminal work by Richard T. T. Forman in *Urban Regions: Ecology and Planning Beyond the City*, 2008.

13 For more details regarding the competitiveness agenda for cities see The World Economic Forum's *The Competitiveness of Cities* at www.weforum.org

14 Sukopp, H. (ed.), 1990, *Urban Ecology: The Example of Berlin.*

15 See 'Why We No Longer Recommend a 40 Percent Urban Tree Canopy', American Forest Association, www.americanforests.org

16 See 'Automated Vehicles Safety', www.nhtsa.gov

CHAPTER 7

1 https://www.reinventingparking.org/2013/02/cars-are-parked-95-of-time-lets-check.html

2 Regarding US latest commuting trend see https://www.usnews.com/opinion/economic-intelligence/articles/2017-09-18/what-new-census-data-reveal-about-american-commuting-patterns

3 See 'The Responsive Network (Part 3/3)', 22 February 2018, https://medium.com/citymapper

4 The most significant frictions and challenges have emerged in relation to the regulatory framework (buses have to run on fixed routes) and public acceptance (residents have to be consulted when buses change routes) rather than the technology, and despite City Mapper's decision to engage in a dialogue with Transport for London. For CAVs these issues risk manifold multiplication of the pushy tactics of start-ups entering the market and disregarding the problems they create, as they see them as necessary costs of implementation.

5 See 'Stories of Cities # 36: How Copenhagen Rejected Modernist "Utopia"' www.theguardian.com/cities/ or read first hand in Jan Gehl's *Life Between Buildings*, 1971.

6 Even the Mayor's Transport Strategy: Draft for Public Consultation (June 2017) relegated CAVs to a cautious open-ended paragraph on page 261.

7 See ENOTRANS 'Adopting and Adapting: States and Automated Vehicles' for the current state of policies related to automated driving in the US, www.enotrans.org

8 www.ons.gov.uk

9 Jeffrey Passel and D'Vera Cohn, 'US Population Projections: 2005–2050', Pew Research Center, 11 February 2008, www.pewresearch.org

Note: page numbers in italics refer to figures.

accesibiility 41, 48
accident 25
adoption 32, 39–40, 88–90, 114–16, 137
advanced traffic management systems (ATMS) 28, 137
air quality 24, 137
artificial intelligence (AI) 137
automation levels 31–2, 137
automobile cities 6–7
autonomous vehicles 28–34
building form 51–2, 86, 114, 120
bus lanes 74, 78
bus networks 120, 132
bus stops 48
buses 74, 78, 132, 135
capacity 38, 44, 137
car ownership 10, 32
car resistance 22–4
carrying capacity 38, 44, 137
CAV cells 74–6, 75
CAV Corridors (C-Corridors) 137
   London case study 71, 75, 75–6, 79
   Los Angeles case study 114–16
charging facilites 39, 76
climate change 24
commuting 34
   see also urban sprawl
Complete Streets 137–8
congestion 39, 61, 119, 138
congestion charges 69, 98–9, 113
connected autonomous vehicles (CAVs) 28, 40, 138
connected vehicle (CV) 138
cost considerations 32, 34–5, 129
   see also income generation
curb (US) *see* kerbside areas
cycle lanes 45, 46, 48–9
   London case study 78, 79, 82
   Los Angeles case study 106, 109
cycling 45, 48–9, 55
   London case study 75, 76, 82
   Los Angeles case study 114
deliveries 49, 52, 122, 126
demarcation of space 45–6, 76, 111–12
demographics 34
demonstration projects *see* trials
design treatments 47
   see also threshold treatments
drones 40, 126
drop-off/pick-up space *see* hop-on/drop-off (HODOs)
ecological considerations 41, 55, 86, 103, 104–5
economics *see* cost considerations
electric vehicles (EV) 40, 138
   see also charging facilites
emergency vehicles 49–50
environmental quality 21–2, 120
   see also air quality; ecological considerations; green space
first and last mile 40, 55, 138
future scenarios 88–90, 127, 130–4
geofencing 138
goods collection and delivery 49, 52, 85, 122, 126
green space
   London case study 82, 84, 86–7
   Los Angeles case study 100, 102, 103–5, 110, 111
growth management 138–9
Healthy Streets 139
heat island effect 104
hop-on/drop-off (HODOs) 39, 44, 46, 51, 139
   London case study 78, 79, 82, 86

Los Angeles case study 106, 112, 113
income generation 86, 122
induced demand 44, 139
kerbs 46
kerbside areas 46, 48, 111–12, 113, 139
   see also hop-on/drop-off (HODOs)
lane sharing 78
legibility 44, 46–7
levels of automation 31–2, 137
local identity *see* place making
local streets *see* place making
logistic centres 52, 85
   see also goods collection and delivery
London case study 67–90
   CAV Cells and Corridors 74–6, 79, 80
   context 61–3
   future scenarios 88–90
   green space 82, 84, 86–7
   implications for the built form 86
   local streets 82–4
   main roads 77–8
   public space 50–1, 86
   public transport 78
   street type relationships 85
Los Angeles case study 91–116
   CAV Corridors (C-Corridors) 114–16
   CAVs supergrid 98–100
   context 63–4
   district movement streets 98, 106, 107
   future scenarios 127, 132, 133, 134
   green space 100, 102, 103–5, 110, 111
   implications for the built form 102–5, 114
   interim phase 114–16
   local intersections 110–11
   local movement streets 108–9
   pavement kerb 111–12
   public space 100, 101, 110, 111
   summary 118
   sustainability and resilience 102–5
macroblock *see* superblock
manually-driven automobiles 49, 71, 85, 114
manufacturers 31
market predication 39–40
mass transit *see* public transport
microtransit 40, 120, 139
mixed-use environments 42, 65, 86, 88
mobility as a service (MaaS) 40, 41, 139
mobility hubs (NEMOHs) 50–1, 140
   London case study 71, 75, 76, 85–6, 87
   Los Angeles case study 98, 99, 104, 106, 112–13
movement hierarchy 49, 64, 98, 113, 116, 122
movement network 48, 64, 139
multi-layered network 48–9, 64, 139
multimodality 40, 140
network approach 48–9, 139
open space *see* green space; public space
parking revenues 122
parking space 12, 39–40, 50, 51, 122
   London case study 86
   Los Angeles case study 96, 113
parking standards 140
parks *see* public parks
pavements 45–6, 106, 109, 140
   see also kerbside areas
pedestrian crossings 47, 106, 109
pedestrian movement 45, 49
pedestrian traffic lights 47
pilot projects *see* trials
place making 23, 42
platooning 45, 47, 140
pollution *see* air quality
prioritization 48–9, 76, 121, 135
   see also movement hierarchy

privately-owned CAVs 32, 41, 48, 64, 82, 99, 106, 113, 130
public parks 52, 55, 86, 102, 103–4
public space
   London case study 50–1, 86
   Los Angeles case study 100, 101, 110, 111
public transport 49, 50, 78, 120, 140
   see also buses; microtransit
real estate 40
ride sharing 41, 102, 119, 130, 140
road and street 140
road hierachy 14–15
service areas 40, 52, 86
service vehicles 48, 49
shared occupancy *see* ride sharing
shared space 45–6, 47–8, 140
   London case study 78, 82
   Los Angeles case study 106, 111
shared vehicles 32, 35, 48, 63, 102, 114
   see also ride sharing
sidewalk *see* pavements
signage 44, 45, 47, 85
   see also design treatments; legibility
smart cities 26, 42, 122, 140–1
Smart Ride 120
social considerations 21–2, 23, 41
space reallocation 38, 44, 48–9, 51–2, 103, 122
spatial considerations 11–14
spatial segregation 21, 63
   see also shared space
speed limits 46, 49, 55
   London case study 76
   Los Angeles case study 99, 106, 108
sprawl *see* urban sprawl
street design 15, 39, 44–6, 137–8
   London case study 82–4
   Los Angeles case study 106–9
   see also kerbside areas; shared space
subscription models 40
suburban contexts 34–5, 122
superblock 49, 99, 100, 105, 141
sustainability 23
sustainable urban drainage systems (SUDS) 79, 82, 141
technologies 29
Tesla 31
threshold treatments 47, 79, 85
time management 46, 48, 88, 112
timeline 32
traffic calming 23
traffic signals 44, 45, 47
transit, mass *see* public transport
transit cities 5–6
transition phase 32, 44, 49, 141
transportation-as-a-service *see* mobility as a service (MaaS)
travel modes
   growth in journey stages on selected modes 61
   passenger throughput for different modes 135
trials 28, 55, 121–2
urban densification 41
urban design 41–2, 43–55, 123, 125–7
   see also street design
urban form 20–1, 120, 130–4
urban identity 22
urban sprawl 12, 14, 120, 141
urban structure 14–15, 120
walkability 22, 41, 55
walking cities 4
wayfinding *see* legibility
Waymo LLC 28
zoning 141

PICTURE CREDITS

Unless noted below, all images by the authors.

Chapter 1
Pages 8 & 9 Roschetzky Photography | Shutterstock.com
Pages 18 &19 Sandor Szmutko | Shutterstock.com

Chapter 2
Page 30 Steve Lagreca / Shutterstock.com
Page 33 B.Forenius / Shutterstock.com
Page 36 & 37 Sebastien DURAND / Shutterstock.com

Chapter 3
Page 53 amc / Alamy Stock Photo

Chapter 4
Page 62 UrbanImages / Alamy Stock Photo

Appendix
Page 128 asiastock / Shutterstock.com

Appendix ii
pag 131, 132, 133, 135 Andres Sevtsuk